Stars and Nebulas

Stars and Nebulas

William J. Kaufmann, III

Department of Physics
San Diego State University

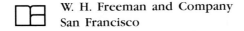 W. H. Freeman and Company
San Francisco

Sponsoring Editor: Arthur C. Bartlett; *Manuscript Editor:* Linda Chaput; *Designer:* Marjorie Spiegelman; *Production Coordinator:* Linda Jupiter; *Illustration Coordinator:* Batyah Janowski; *Artist:* Dale Johnson; *Compositor:* Graphic Typesetting Service; *Printer and Binder:* The Maple-Vail Book Manufacturing Group.

Library of Congress Cataloging in Publication Data

Kaufmann, William J.
     Stars and nebulas.

     Bibliography: p.
     Includes index.
     1. Stars. 2. Nebulae. I. Title.
QB801.6.K38   523.8   78-17544
ISBN 0-7167-0081-6
ISBN 0-7167-0085-9 pbk.

Printed in the United States of America

9   8   7   6   5   4

Cover photograph: The nebula M 16 (also called NGC 6611) in the constellation of Serpens. (Copyright © 1965 by California Institute of Technology and Carnegie Institution of Washington.)

to Louise N. with love

Contents

Preface

The splendor of the night sky has fascinated people for thousands of years. Like countless generations before us, we all have gazed up at the stars and wondered at what we saw. For most of us, this experience is just a fleeting glance accompanied by a few vaguely formulated questions—nothing more than a momentary distraction from the joys and frustrations of daily life. But over the ages some people have paused to ponder the meaning of the stars. The daily rising and setting of the sun, or the silver moon going through its phases every 28 days, implies order rather than chaos to the rational human mind. The desire to fathom this celestial order, to discover the fundamental laws of the universe, is at the heart of astronomy.

In ancient times the workings of the heavens were explained largely by myth and fantasy. People believed that the stars and planets were populated with gods, heroes, demons, and monsters that ruled everything from eclipses to constellations. But then, only a few hundred years ago, a remarkable revelation profoundly—and permanently—changed the course of humanity. It was discovered that the laws that govern our world are the same laws that dictate the behavior and properties of the planets, the stars, and the universe. The force that holds us firmly on the ground is the same force that keeps the earth in orbit about the sun. The chemicals of which rocks and people are made are the same substances of which planets, stars, and galaxies are composed. Thus, by doing experiments in the laboratory we can discover fundamental truths that control the cosmos.

Preface

And by observing the universe we can discover the basic principles at work in the world around us.

In modern astronomy and astrophysics we truly see our oneness with the universe. The laws of the microcosm are the laws of the macrocosm. By exploring the stars and nebulas we can more fully understand and appreciate ourselves and our world.

May 1978   W. J. K.

Stars and Nebulas

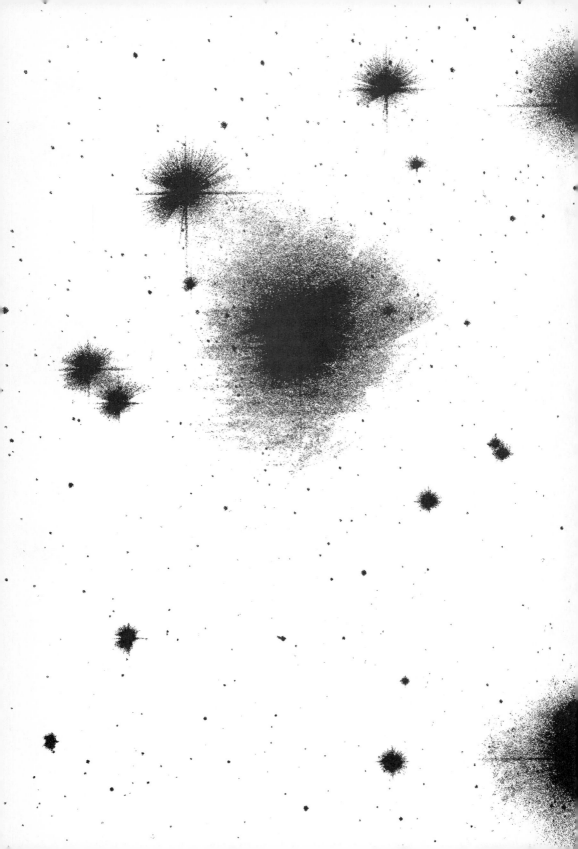

1

A Stellar Scenario

We shall speak of things we cannot understand. We shall discuss concepts we cannot grasp. We shall examine processes we cannot comprehend.

This has become the legacy of modern astronomy and astrophysics. The human intellect has driven us into an unearthly realm far beyond the domain of human experiences.

Our world of human experiences is measured in millimeters and miles, in inches and kilometers. Yet we shall be compelled to speak in dozens of light years as we discuss the distances between the stars.

On our planet, ordinary rocks are among the densest objects we encounter. Yet we shall be challenged to imagine conditions inside white dwarfs and neutron stars where atoms are so densely packed that they are crushed beyond all recognition.

Most of us know what it feels like to hold an ice cube or to be dashed by accident with scalding water. This range encompasses our direct experience of temperature. Yet we shall discover temperatures only a few degrees above absolute zero in the vast interstellar clouds where stars are born. We shall find temperatures of hundreds of millions of degrees at the centers of stars where atoms and elements are forged in thermonuclear fires.

Earthquakes, thunderstorms, hurricanes, and tornadoes are surely the most violent events on earth. Yet we shall see them as tame compared to the cataclysm of a supernova, in which a dying star is torn apart.

To learn about modern astronomy you must want to understand things that are far beyond human experience. Instead of inches or miles, you shall speak in light years and parsecs. Instead of hours or days, you shall think in millions and billions of years. The inconceivable shall become commonplace. You must learn to imagine the unimaginable as you look towards the heavens.

As we gaze out into the starfilled sky on a clear moonless night, it sometimes seems as if we can see millions upon millions of stars. Actually, this is a severe exaggeration: the naked human eye can see a *total* of only 6,000 stars. At any one moment, we see roughly 3,000 stars across the sky; the other 3,000 are below the horizon and thus are hidden from view. Of course, there really are countless

Figure 1-1.   A Cluster of Galaxies
Countless trillions of galaxies are scattered across the universe. A typical galaxy measures 100,000 light years in diameter and contains over 100 billion stars. The vastness of what we see in the heavens makes astronomy one of the most consciousness-expanding fields of human knowledge. (Hale Observatories.)

Figure 1-2. Stonehenge
Fascination and preoccupation with the heavens predate the dawn of recorded history. Monuments such as Stonehenge stand as mute testimony to our ancestors' careful and patient observations of the skies. (Courtesy of the Controller of Her Britannic Majesty's Stationery Office. British Crown Copyright.)

billions of stars in the universe. But aside from the 6,000 stars visible to the naked eye, all the rest are so dim and distant that powerful telescopes or time exposure photography must be used to reveal them.

People have been looking up at the heavens for thousands of years. From the pyramids to Palomar, humanity has searched the skies for clues to its origin and destiny. Many concepts in modern astronomy bear witness to this ancient heritage. For example, our ancestors felt that certain groups of stars suggested figures of animals, heroes, and mythical creatures. These groupings of stars are called *constellations,* and most were given names from mythology and folklore. There are animals such as Leo (the lion), Taurus (the bull), Aquila (the eagle), and Cygnus (the swan). There are heroes from mythology such as Hercules, Perseus, and Orion. And then there are imaginary creatures such as Capricornus (the sea goat), Draco (the dragon), Cetus (the sea monster), and Pegasus (the winged horse).

There are 88 constellations that cover the entire sky. Even today, astronomers often find it convenient to use them for denoting certain regions in the sky. Just as someone might speak of a particular city in France, the astronomer would speak of a particular galaxy in Virgo. Just as you might read of an earthquake in China, the astronomer might discover a comet in Cassiopeia.

Just as most of the names of constellations come to us from antiquity, so do most of the names of the brighter stars. But whereas the constellations are rooted in ancient Greek and Roman mythology, the vast majority of star names are Islamic and Arabic. During the Dark Ages, when western civilization in Europe was largely in disarray, Islamic cultures reached their peak. It was at this time that Moslem astronomers named most of the bright stars. For example, the seven stars in the Big Dipper are: Dubhe, Merak, Phecda, Mergrez, Alioth, Mizar, and Alkaid.

These star names are poetic and romantic, but often they prove cumbersome and inconvenient. For example, the brightest star in Libra (the scales) is Zubenelgenubi. Few astronomers feel inspired to memorize long lists of Arabic names; even fewer can spell. And so, in the 1600s, a new naming system was invented. This system labels

7

a

Figure 1-3a. Scorpius
Scorpius, the scorpion, is a prominent constellation of the summer sky. Modern astronomers still use these ancient constellations to denote certain regions of the sky. (Courtesy of Sara W. Hoffman.)

Figure 1-3b. A Drawing of Scorpius
Star charts from old astronomy books were lavishly illustrated. Although these charts were works of art, they contained many inconsistencies concerning the locations and boundaries of constellations. (Griffith Observatory.)

Figure 1-3c. The Modern Scorpius
In 1928, astronomers from around the world decided upon the boundaries and locations of all 88 constellations. As shown here, the boundaries generally run east to west and north to south.

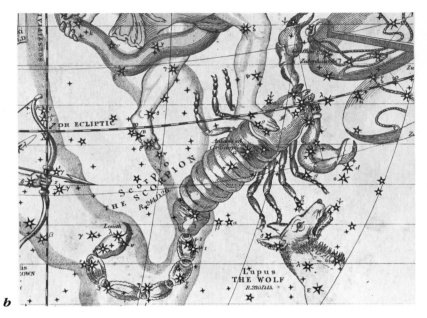

b

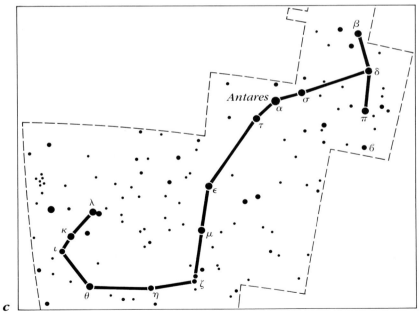

c

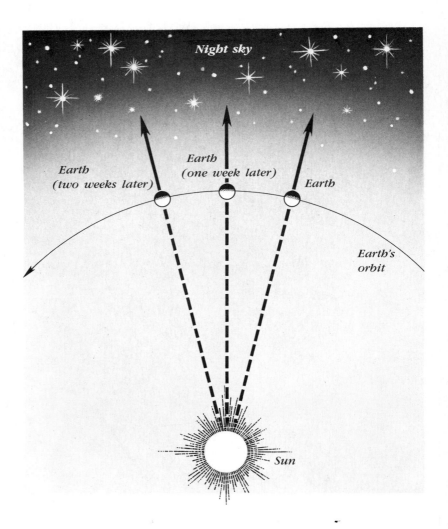

Night sky

*Earth
(two weeks later)*

*Earth
(one week later)*

Earth

*Earth's
orbit*

Sun

Figure 1-4. The Changing Night Sky

*As the earth moves in its orbit around the sun, the nighttime side
of the earth moves to face different parts of the heavens. This is
why different constellations are seen in the sky at different times of
the year.*

stars in a constellation with letters from the Greek alphabet (alpha, beta, gamma, delta, epsilon, and so on). The brightest star in a constellation is usually called α, the second brightest is β, and the third brightest is γ. Thus, for example, the brightest star in Leo (the lion) is α Lionis. Zubenelgenubi is α Librae. And the third brightest star in Orion (the hunter) is γ Orionis. All this seems to make life a little easier for the astronomer.

The earth rotates once every twenty-four hours. This is why the sun, moon, planets, and stars appear to rise in the east and set in the west. Thus, during the course of a night, the sky changes. New stars and constellations rise in the east while old ones set in the west. But in addition, the sky gradually changes from night to night. The reason for this is that the earth is orbiting the sun. The earth takes a full year (365¼ days) to go once around the sun. As the earth moves along its orbit, the dark, nighttime side of the earth gradually faces different parts of the heavens, as shown in Figure 1-4. This effect is difficult to notice from one night to the next. At 10 p.m. on April 2 the sky looks almost exactly the same as at 10 p.m. on April 3. But over weeks and months, the changes are very noticeable: the sky in April looks very different from the sky in August. A series of twelve star charts (one for each month) at the end of this book gives the appearance of the sky during the evening hours over the course of the year.

Even the most casual glance at the nighttime sky shows that there are bright stars and there are dim stars. But appearances can be deceiving. A star that looks bright in the nighttime sky might really be a very dim star that happens to be nearby. Conversely, a very faint-appearing star in the sky might really be a very luminous star that just happens to be extremely far away. In short, a glance at the sky does not reveal anything about the *true* nature of the stars themselves.

Whereas casual observers can derive pleasure from the stars' appearance, astronomers and astrophysicists would prefer to know their *true* brightnesses rather than just their *apparent* brightnesses. The true brightness of a star would tell scientists how much power (for example, in kilowatts) the star actually emits from its surface in the form of starlight. But in order to discover the true brightnesses of stars, it is first necessary to know how far away they are.

Figure 1-5. The Hale Telescope
*This dome on Palomar Mountain houses one of the largest
telescopes in the world. The telescope's primary mirror is 200 inches
(nearly 17 feet) in diameter. Looking through this telescope,
astronomers can see stars that are 400,000 times fainter than the
dimmest stars visible to the naked eye. (Hale Observatories.)*

Imagine walking down the street. As you move from one location to another, nearby objects appear to shift their positions with respect to the background scenery, as shown in Figure 1-6. This phenomenon, called *parallax,* is part of our everyday common experience. Similarly, as the earth goes around the sun, nearby stars appear to shift their positions with respect to the distant background stars, as shown in Figure 1-7. But the stars (even the nearest ones) are so far away that their parallaxes are exceedingly difficult to detect. Nevertheless, if this parallactic shifting could be measured, the simplest surveying formulas from trigonometry would immediately give stellar distances.

By the early nineteenth century, astronomers had finally developed the necessary techniques to measure stellar parallaxes. By observing the tiny displacements of nearby stars over many months, astronomers finally had the opportunity to discover where the stars really are.

A dim star called 61 Cygni was the first to have its parallax measured. This was soon followed by the bright star Vega and then α Centauri. Since these beginnings, which took place in the 1840s, the parallaxes of over a thousand stars have been measured with reasonable accuracy. In all cases, the parallactic angle — the angle through which the star appears to shift as the earth moves about the sun — is extremely small. This, in turn, means that these stars must be extremely far away. Indeed, these "nearby" stars are trillions upon trillions of miles away.

As you might expect, expressing distances to the stars in trillions of miles (or kilometers) is very cumbersome. Indeed, it would be just as awkward as talking about distances between cities on the earth in inches or millimeters. No one likes to think of the distance between New York and Los Angeles as 155,000,000 inches. Similarly, astronomers would not find it helpful to express the distance to Vega as 15,600,000,000,000 miles. To cope with this difficulty, the *light year* was invented. A light year is simply how far light travels in one year: about 6 trillion miles. Using the light year, we find that the distances to the stars can be expressed with very convenient numbers. For example, Vega is 26 light years away. Sirius, the brightest-appearing star in the sky, is 9 light years away. The nearest star, Proxima Centauri, is only 4¼ light years away. Most of the stars we

13

Figure 1-6. Parallax
Imagine looking at some nearby object (a tree) seen against a distant background (mountains). If you move from one location to another, the nearby object will appear to shift with respect to the distant background scenery. This phenomenon is called parallax, and it is part of our everyday experience.

14

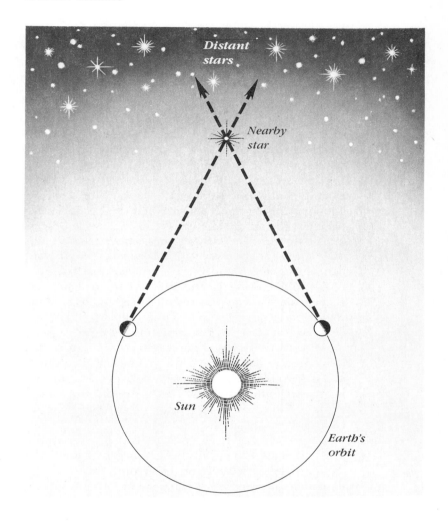

Figure 1-7. Stellar Parallax
*Since the earth goes around the sun, nearby stars appear to change
their positions with respect to the background stars. This
phenomenon is very difficult to measure because all the stars (even
the nearest ones) are extremely far away.*

see with our naked eyes at night are within 1,000 light years of the earth.

Once the distance to a star is known, it is an easy matter to calculate its true brightness. For example, consider Polaris, the North Pole star. Polaris, at the end of the handle of the Little Dipper, is not an especially outstanding star in the nighttime sky. But Polaris is extremely far away. The distance to Polaris is 680 light years. Since Polaris is so remote and yet can be easily seen with the naked eye, the star must be exceedingly luminous. From knowing the apparent brightness (accurately measured, for example, with sensitive photoelectric equipment mounted at the focus of a telescope) and from knowing the distance (determined, for example, from measuring parallax), the astronomer can calculate the true brightness of the star. Thus, we find that Polaris shines with a true brightness of 10,000 suns. A true brightness of 10,000 suns is the *only* luminosity that is compatible with both the star's distance and its apparent brightness.

In discussing the true brightnesses, or *luminosities,* of stars, it is convenient to compare them to the luminosity of the sun. Our sun shines with a power output of 400 trillion trillion watts. For convenience, we say that this amount of power equals 1 *sun.* A star that is intrinsically two times brighter than the sun is said to have a luminosity of 2 suns. A star that has only one-half the power output of the sun is said to have a luminosity of ½ sun. In other words, the sun's luminosity is the standard. The luminosities of other stars are then expressed in terms of how many times brighter or dimmer they are compared to the sun.

The luminosities of stars cover an enormous range. The brightest stars shine with a brilliance of a million suns. And the dimmest stars emit only one-millionth of a sun. It is interesting to note that our sun lies in the middle of the full range of stellar luminosities. This is our first clue that we are in orbit around a very ordinary star.

Only a few hundred years ago it was generally believed that the stars were so incredibly remote that we had no hope of ever discovering their true nature. It was felt that these tiny, inaccessible pinpoints of light in the nighttime sky would forever be beyond our reach and understanding. But at approximately the same time that astronomers first began measuring the distances to the stars, physicists were making important discoveries about the properties of light itself. These revelations showed that a beam of light can contain an enormous amount of information. Armed with this new knowledge, astronomers realized that they could dissect and analyze light from the stars. In doing so, they hoped to unlock the secrets contained in starlight and thereby discover the true nature of what they see in the sky.

A modern understanding of the nature of light began with the pioneering work of Sir Isaac Newton. In the mid-1660s, Newton made the remarkable discovery that when a beam of white light is passed through a glass prism it breaks into the colors of the rainbow. White light goes into the prism, and a rainbow of colors, called a *spectrum,* comes out. This demonstrates that white light actually consists of all the colors of the rainbow combined. This simple but critical discovery showed how light could be dissected into its basic parts.

The nature of these basic parts was revealed with a simple experiment performed by the English physicist Thomas Young. Imagine shining a beam of light onto two closely-spaced slits in a sheet of cardboard or metal, as shown in Figure 1-9. Furthermore, suppose that only light of a single color is used, rather than white light, which is all colors combined. It was generally believed that the light should go straight through the slits and produce two bright lines on the screen beyond. But when Young performed this experiment, he was surprised to find that a whole host of dark and light bands appeared on the screen.

Prior to Young's experiment in 1801, most scientists believed that light consisted of tiny packets of energy. From arguments proposed by Newton, a beam of light was thought of as a stream of tiny billiard balls. That is why everyone expected to see only *two* bright lines on Young's screen. Since there were only two slits in Young's apparatus, only two beams of light could pass through.

But Young found that a series of light and dark bands, called *interference fringes,* appeared on the screen beyond the two slits.

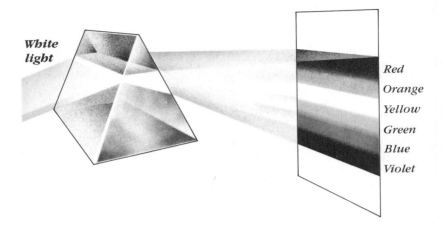

Figure 1-8. A Prism and a Spectrum
When a beam of white light is passed through a prism, the light is broken up into the colors of the rainbow. This experiment, first performed by Newton around 1665, proves that white light actually consists of all colors.

This meant that there was something very wrong with the mechanical, billiard-ball ideas about light.

It was soon realized that the behavior of light in Young's experiments was remarkably similar to the behavior of water waves in the ocean. Imagine water waves in the ocean pounding against a reef or barrier that has two openings. The pattern of ripples that strike the shore beyond the breakwater are essentially the same (but on a much larger scale) as the pattern of interference fringes that appeared on Young's screen. Young's experiment therefore constituted the first direct proof that light consists of *waves*.

More than a century had elapsed between Newton's experiment with a prism and Young's discovery of the wave nature of light.

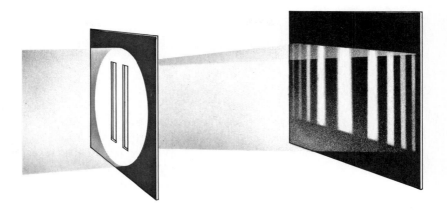

Figure 1-9.   The Double Slit Experiment
When a coherent beam of monochromatic light (that is, light of one color) is shone on a pair of closely spaced slits, interference fringes appear on the screen beyond. This crucial experiment proves that light consists of waves.

The cause of the delay was rooted in the fact that the wavelengths of visible light are extremely small. As diagramed in Figure 1-10, the distance between successive crests or valleys in a wave is called the *wavelength*. The wavelengths of visible light range from 0.000016-inch for violet light to 0.000028-inch for red light. Intermediate colors of the spectrum have intermediate wavelengths.

Of course, the inch is not a convenient unit of measure with which to express wavelengths of light. Most scientists prefer to use angstroms (abbreviated Å). There are 10 billion angstroms in a meter or, alternatively, about one-quarter billion angstroms in an inch. Thus, visible light covers the wavelength range from about 4,000 Å to 7,400 Å.

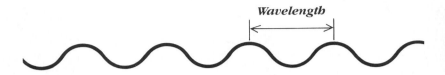

Figure 1-10.   Waves and Wavelength
Light is a wave phenomenon. The wavelength of a particular
color of light is the distance between successive crests (or troughs)
in the wave. Wavelengths of visible light range from 4,000 Å (for
violet light) to about 7,400 Å (for red light).

During the late 1800s, many physicists began examining the
possibility that radiations exist outside the range of visible light. For
example, they wondered what exists at wavelengths shorter than
4,000 Å. Or at wavelengths longer than 7,400 Å. And soon one discovery after the next began to pour in.

In 1887, Heinrich Hertz used electric sparks to produce an
invisible form of light having wavelengths of roughly one foot. Today
these radiations are called radio waves. By 1895, Wilhelm K. Röntgen
invented an apparatus that produced another form of invisible light
having wavelengths of approximately 100 Å. Today these waves are
called X rays.

Gradually all the pieces began to fit together. By the turn of
the twentieth century it was becoming clear that X rays, γ rays, radio
waves, ultraviolet, and infrared radiation, as well as ordinary visible
light, are really the same thing. Together they form the *electromagne-*
tic spectrum. Today we realize that the visible light to which our eyes
respond is only a small portion of an immense continuous range of
radiations. The visible light with which we view the universe is only
a tiny fraction of all the wavelengths of radiation that are coming to
us from the depths of space.

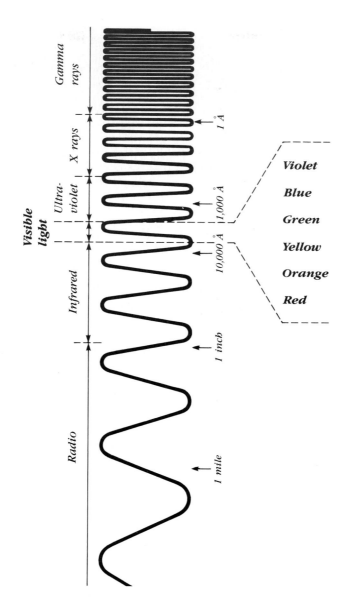

Figure 1-11. The Electromagnetic Spectrum
The electromagnetic spectrum is the complete array of types of radiation. Notice that ordinary visible light constitutes only a small portion of the entire range.

Figure 1-12. The Solar Spectrum
*The sun's spectrum contains over 15,000 spectral lines. A small
portion of the solar spectrum, around 5,000 Å, is shown here. These*

As long as astronomers observe the sky only at visible
wavelengths, they miss most of the story. All the information contained at nonvisible wavelengths from the heavens is lost. But during
the nineteenth century, no one knew how to detect or "see" anything
except visible light. So they were determined to make the most of it.
A lot of effort therefore went into building better telescopes. In
addition to improving observations, there was an occasional bonanza
as some new property of light was uncovered.

In 1814, the gifted German optician Joseph Fraunhofer invented a method of precisely determining the angles through which
different colors of light are bent by various kinds of glass. Ideally this
technique would be used for testing various types of glass before
grinding lenses. His method was based on a discovery by William
Wollaston in 1802 that the spectrum of sunlight contained several
dark lines. Fraunhofer realized that these *spectral lines* could serve as
ideal reference markers in his experiments with glass.

wavelengths encompass the blue part of the sun's spectrum. (Hale Observatories.)

Fraunhofer promptly set about the business of repeating Newton's classic experiment with a beam of sunlight and a prism, except that this time the resulting spectrum would be greatly magnified. Much to everyone's surprise, the solar spectrum contained hundreds upon hundreds of fine, irregularly spaced dark lines. Fraunhofer counted 700 such lines and today we know of over 15,000.

A small portion of the sun's spectrum is shown in Figure 1-12, which contains several thousand lines. Where do these lines come from? What do they mean? What causes them? These questions remained totally unanswered for many decades. The final solution to the mystery would be found only when we came to know the detailed structure of atoms and the chemical composition of stars.

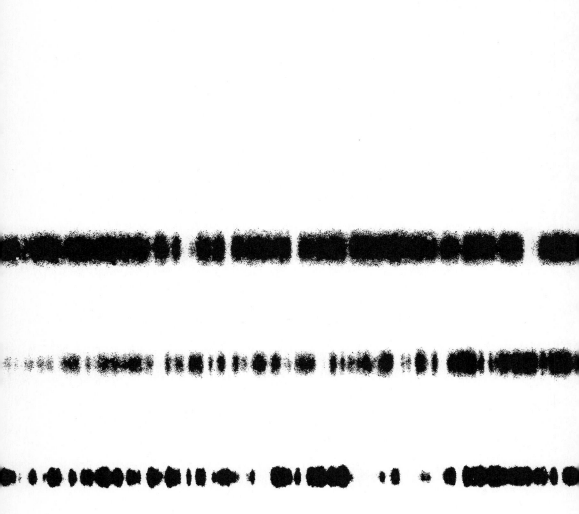

2

The Laws of Light

Almost everything we know about the universe beyond Earth comes to us through light. We can examine meteorites that have fallen from the sky and the few hundred pounds of rocks that have been brought back from the moon, but light alone reveals countless stars, nebulas, and galaxies scattered across the cosmos.

As long as we do not understand light, astronomy is restricted to peering through telescopes and seeing pretty things. As long as we do not know how to dissect and analyze radiation from the stars, we shall never discover the true nature of what we observe.

One of the greatest achievements of nineteenth-century science was the discovery of the properties of light. It began in 1801 when we learned that we can speak of light *waves* (Figure 1-9), and it took a full century to develop a complete explanation of the spectral lines that Fraunhofer discovered in sunlight (Figure 1-12). But once we had this understanding, astronomy would flourish as never before. The labors of nineteenth century physicists directly culminated in the revelations of twentieth century astronomers.

Profound insight into the nature of light began with an attempt to understand ordinary phenomena. Imagine a simple experiment in which you take a bar of iron and heat it, perhaps with the aid of a blowtorch. Initially, the bar of iron will start glowing with a deep dull reddish color. But as the temperature of the bar rises, it glows with a brighter, reddish-orange light. At still higher temperatures, the bar shines with a brilliant yellowish-white light. And if the bar could be prevented from melting and vaporizing, at very high temperatures it would gleam with a dazzling blue-white light.

This simple experiment demonstrates two very important physical principles. First of all, the total amount of energy radiated by a hot object depends on the temperature of the object. The hotter the object, the more light it gives off. But in addition, the dominant color of light emitted also depends on the object's temperature. Relatively cool objects primarily emit long-wavelength radiation while hotter objects emit much shorter wavelength radiation.

In 1879, Josef Stefan performed a series of experiments in which he measured the total energy emitted by objects at various temperatures. As a result of his experiments, Stefan was able to formulate a simple but precise mathematical relationship between the temperature of an object and the total amount of radiation it emits. This relationship is today called *Stefan's Law.*

A few years later, in 1893, Wilhelm Wien succeeded in discovering a very simple relationship between the temperature of an object and the wavelength at which the object emits the most light. Of course, a glowing object at a specific temperature gives off a wide range of wavelengths. But this relationship, called *Wien's Law,* tells us precisely at what wavelength the most radiation is emitted.

Stefan's law and Wien's law together describe some of the basic phenomena associated with radiation from hot, glowing objects — such as the bar of iron discussed earlier. Actually, physicists realized that the results of their experiments depended slightly on whether they used bars of iron, or silver, or carbon, or whatever. To circumvent this difficulty, the concept of a *blackbody* was invented. A blackbody is a "perfect radiator" in the sense that the radiation emitted by a blackbody does not depend on extraneous details such as its shape, size, or chemical composition.

The radiation emitted by an ideal blackbody is naturally called *blackbody radiation.* A real object at a specific temperature emits certain amounts of radiation at various wavelengths that often approximate blackbody radiation. For example, Figure 2-1 shows how much energy is radiated from the sun at various wavelengths. The dashed line on the graph is the curve for a blackbody at 5,800° Kelvin (K). We can therefore say that the surface temperature of the sun is roughly 5,800°K. Of course, the sun is not an ideal blackbody; the wiggles in the sun's energy curve are due to the effects of chemicals and conditions in the sun.

If you consider these early investigations about radiation from hot objects, it is easy to understand one reason why scientists so strongly prefer to speak of temperatures on the Kelvin scale. That scale alone is directly related to fundamental physical processes of matter and radiation. The Kelvin scale measures temperature upward from absolute zero (0°K), the coldest possible temperature. According to Stefan's Law, everything above absolute zero emits some form of radiation. For example, according to Wien's Law, an object a few degrees above absolute zero emits primarily radio waves, an object a few thousand degrees above absolute zero emits mostly visible light,

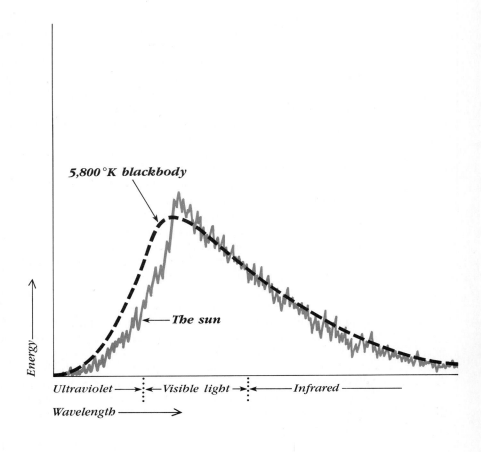

Figure 2-1. The Sun and a Blackbody
The jagged curve shows how much energy is emitted from the sun's surface at various wavelengths. The dashed curve is for a blackbody at 5,800 °K. The sun's energy distribution does not follow a blackbody curve exactly because of the effects of chemicals and conditions in the solar atmosphere.

and at a few million degrees lots of X rays are given off. Only at absolute zero is an object so cold that it emits no radiation at all. In contrast, the familiar Fahrenheit scale was based on the temperature of the human body (believed to be 100°F), and the Celcius scale was based on the melting and boiling points of water (0°C and 100°C, respectively).

Although the work of Stefan and Wien constituted new major advances, it was soon painfully clear that physicists faced some big problems. Stefan's Law and Wien's Law describe only two details about the radiation emitted by glowing objects. The overall picture about the amount of radiation emitted at *all* wavelengths for a particular temperature soon became even more mysterious than ever.

Figure 2-3 illustrates the distribution of energy radiated from glowing objects at various temperatures. The curves, called blackbody curves, clearly display both Stefan's Law and Wien's Law. A cool object has a low curve that peaks at long wavelengths, whereas a hotter object has a much higher curve that peaks at a shorter wavelength. But why do these curves have the characteristic shapes given in Figure 2-3?

In order to understand the details of how objects emit light, one must account completely for the characteristic shapes of blackbody curves. These curves, which were obtained by careful laboratory experiments, baffled scientists during the late 1800s. In fact, science of the nineteenth century was at a complete loss to explain the details of how hot objects glow. As long as you think only of waves, it is impossible to understand exactly how a hot object radiates light.

The breakthrough came during the first decade of the twentieth century. Before this time, it was believed that the energy could be absorbed or emitted from an object in any quantity whatsoever. In other words, a hot object would emit waves of radiation in a smooth and continuous fashion. But this approach failed to explain the details of blackbody radiation. In desperation, the German physicist Max Planck considered something that seemed preposterous: energy might be emitted (or absorbed) only in tiny discrete amounts, rather than in a continuous fashion. After all, by that time, it was known that matter really consists of very tiny pieces called atoms. So, perhaps energy comes in very tiny pieces also.

29

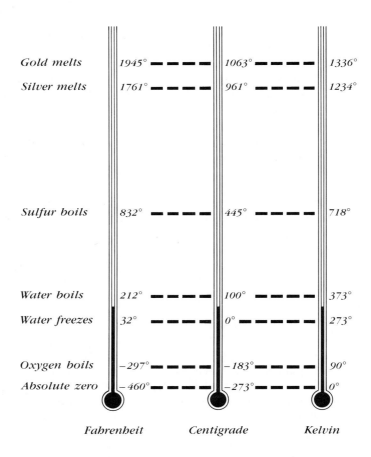

Figure 2-2. **Temperature Scales**
Scientists prefer to use the Kelvin scale when discussing temperatures. Unlike other commonly used scales, the Kelvin scale is rooted in fundamental physics. This illustration relates the Kelvin, Centigrade, and Fahrenheit scales.

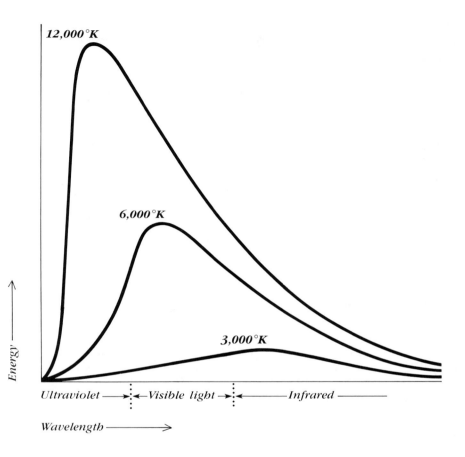

Figure 2-3. Blackbody Curves
Each curve illustrates the amount of energy radiated at various wavelengths from a hot object at a specific temperature. Using classical, nineteenth century ideas about light, it was flatly impossible to explain the shapes of these curves.

In developing his ideas, Planck made the critical discovery that *the amount of energy carried by light depends on the wavelength of the light.* The shorter the wavelength, the higher the energy. For example, blue light contains more energy than red light.

Planck's bold and revolutionary inspiration met with immediate success. Assuming that energy is emitted and absorbed in very small discrete quantities (later called photons), Planck was able to derive a mathematical formula that completely described blackbody curves such as those shown in Figure 2-3. The derivation of this one equation marked a major turning point in physics. From that point on, mankind would view the structure of physical reality from an entirely new perspective — the perspective of quantum mechanics.

The quantum theory of light was invented to explain exactly how and why hot objects emit radiation. In order to understand why a horseshoe in a blacksmith's forge glows with a reddish light, it was necessary to discover that radiation comes in very tiny discrete packets. This profound insight into the nature of light would soon be used to discover the structure of atoms and to explain the mysterious spectral lines that Fraunhofer had discovered in sunlight.

Fraunhofer's discovery of numerous dark lines in the sun's spectrum lay dormant for almost half a century. Nobody knew what these lines meant or why they were there. But science often evolves in unexpected ways. The first clues to the meaning of the sun's spectrum would come from the invention of the Bunsen burner.

In 1856, the German chemist Robert W. E. Bunsen invented an improved gas burner. This burner had the novel advantage that it produced a clean colorless flame. Thus, when chemicals were sprinkled in the flame, the colors that were produced were caused solely by the burning chemical substances and not by the flame itself.

Upon hearing of this invention, one of Bunsen's colleagues at the University of Heidelberg, Gustav Kirchhoff, proposed that the colors from burning chemicals could be more easily studied with a prism. Bunsen and Kirchhoff therefore collaborated to construct a spectroscope, a device consisting of a prism and several lenses, by which the spectra of chemicals might be magnified and examined. They promptly discovered that *each chemical element produces its own characteristic pattern of spectral lines.*

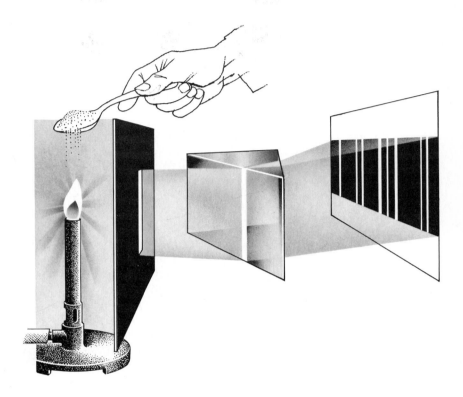

Figure 2-4. The Kirchhoff-Bunsen Experiment
In the mid-1850s, Kirchhoff and Bunsen discovered that when a chemical substance is heated and vaporized, the resulting spectrum exhibits a series of bright spectral lines. In addition they found that each chemical element produces its own characteristic pattern of spectral lines.

Figure 2-5. Iron on the Sun
The upper spectrum is a portion of the blue part of the sun's spectrum (4,200 to 4,300 Å). Numerous dark absorption lines are seen against the blue colors of the rainbow. The lower spectrum is a corresponding portion of the spectrum of vaporized

In the early nineteenth century it was generally assumed that mankind would never know what the stars are made of. The pioneering work of Bunsen and Kirchhoff permanently dispelled this erroneous belief. The meaning of Fraunhofer's lines in the solar spectrum was now becoming clear: They were caused by chemicals present in the inferno of the sun's atmosphere. Spectral analysis had been born.

The central idea behind spectral analysis is very simple. Take an element, place it in the flame of a Bunsen burner, view the resulting light through a spectroscope, and carefully record the pattern of spectral lines. This pattern is the unmistakable, indelible fingerprint of that particular chemical. If that same pattern of lines exists in the spectrum of the sun or a star, that chemical must be present in its atmosphere. In this way an analysis of starlight begins to yield secrets of the stars themselves.

The invention of spectral analysis also resulted in the discovery of new elements. For example, during the course of their experiments, Bunsen and Kirchhoff found some unfamiliar spectral lines in the spectrum of mineral water vapor. They soon determined that these new lines were caused by two previously unknown elements: rubidium and cesium.

34

iron in an iron arc. Several bright emission lines are seen against a black background. The fact that the bright iron lines coincide with some of the dark solar lines proves that there is some iron (albeit a very tiny amount) in the sun's atmosphere. (Hale Observatories.)

It was also found that a line in the green part of the solar spectrum did not correspond to any known substance here on earth. It was therefore concluded that a chemical element, unknown here on earth, must exist on the sun. This element was called helium, from the Greek word *helios,* which means "sun." Helium was discovered in the solar atmosphere before it was known on our planet.

By the early 1860s, Kirchhoff's experiments had progressed to the point that he could formulate three general statements about spectra. The first of these *laws of spectral analysis* simply states that a hot glowing object emits a continuous spectrum. In other words, if you pass the light from a hot object (such as the glowing filament of a light bulb) through a prism, you will see a complete rainbow of colors without any spectral lines.

Kirchhoff's second law states that when a source of a continuous spectrum is viewed through some gas or vapor, then dark spectral lines are seen in the resulting spectrum. Conversely, according to the remaining law, if this same gas or vapor is viewed at an oblique angle away from the source of white light, then a series of bright spectral lines is seen against an otherwise black background. The remarkable fact is that for a particular chemical the bright spectral lines (called *emission lines*) occur in exactly the same sequence and at exactly the same wavelengths as the dark spectral lines (called *absorption lines*). Indeed, it is as if the chemicals extract light at specific wavelengths from the continuous rainbow, thereby producing an *absorption line spectrum*. But these chemicals then reemit this extracted light in all directions and produce an *emission line spectrum* viewed at oblique angles, as diagrammed in Figure 2-6.

But what does this mean? Why is it that iron vapor shows about 6,000 lines across the visible spectrum whereas sodium vapor has only two lines in the yellow portion of the spectrum? How are we to understand the various patterns of spectral lines, and what exactly does a chemical do to form a spectral line in the first place? Answers to these questions were beyond the scope of nineteenth century physics.

In 1910, Lord Ernest Rutherford and his colleagues were experimenting with a recently discovered phenomenon called radioactivity. Radioactivity is a phenomenon whereby certain elements such as radium and uranium emit particles. One type of subatomic particle — an α-particle — is emitted from radioactive substances with considerable energy.

Rutherford and his colleagues were examining how α-particles interacted with various pieces of matter. In particular, they directed a beam of α-particles against a thin sheet of metal. Most of the α-particles easily passed directly through the thin sheet with almost no deflection at all. This is what everyone expected. But after only a couple of days of experimenting, they discovered that a few α-particles were reflected backwards. In other words, while the vast majority of particles passed nearly unhampered through the metal sheet, occasionally one hit something inside the sheet that caused the α-particle to rebound.

The profound implications of this experiment were instantly clear to Rutherford. For many years, scientists had been talking of *atoms* as the fundamental building blocks of all matter. Tiny, negatively charged particles called *electrons* were known to be present in atoms, but Rutherford's experiment showed how the atom is organized. Atoms are mostly empty space; after all, most of the α-particles easily penetrate the metal foil. But the occasional recoil of

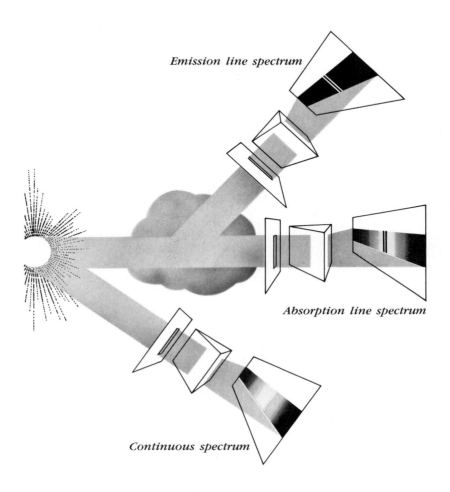

Emission line spectrum

Absorption line spectrum

Continuous spectrum

Figure 2-6. Kirchhoff's Laws
A hot glowing object emits a continuous spectrum. If this white light is passed through some cool gas, then dark absorption lines appear in the resulting spectrum. And if this same gas is examined at an oblique angle, bright emission lines are seen.

37

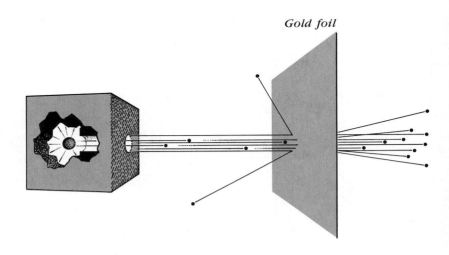

Gold foil

Figure 2-7.   Rutherford's Experiment
Alpha particles from a radioactive source are directed against a thin metal foil. Most α-particles pass through the foil with very little deflection. Occasionally, however, an α-particle recoils dramatically, which indicates that it has collided with the nucleus of an atom. This crucial experiment proved that atoms have nuclei.

α-particles proved that there must be a very compact, extremely dense lump of matter at the heart of each atom. Rutherford therefore proposed that atoms really look like miniature solar systems. Most of the matter of an atom resides in its *nucleus*. Tiny electrons presumably orbit the massive nucleus just as tiny planets orbit the massive sun. Since the electrons are negatively charged, the nucleus must carry a positive charge so that the whole atom is electrically neutral. The atom is therefore held together by electric forces (recall the adage "opposites attract") just as the solar system is held together by gravity.

Nothing seemed more absurd than Rutherford's proposal. After all, why shouldn't the negatively charged electrons plunge into the positively charged nuclei? Every ounce of classical physics clearly predicted that Rutherford's atoms, if they existed, would collapse very rapidly.

A young Danish physicist, Niels Bohr, came to Rutherford's rescue with a direct application of the newly formulated quantum theory of light. It was becoming increasingly clear that energy and light are not continuous phenomena. Instead, energy and light are *quantized:* they come in discrete quantities or pieces called photons. And it occurred to Bohr that perhaps the orbits around the nucleus are also quantized. Instead of a continuous range of possible orbits, perhaps there are only certain orbits for electrons at certain prescribed distances from the nucleus. If this were so, then the electrons could not spiral into the nucleus and Rutherford's atoms would not collapse.

Bohr's ideas, first published in 1913, met with instant success. Not only was Rutherford's model of the atom saved, but all the work of Fraunhofer, Kirchhoff, and Bunsen was promptly understood.

The central idea is that there are only certain allowed orbits for electrons in an atom. In order for an electron to go from a low orbit near the nucleus to a higher orbit, the atom must absorb a very specific amount of energy. Similarly, in order to drop from a high orbit down to a low orbit, the atom releases a very precise amount of energy. Electrons will not budge either up or down in an atom unless a very precise amount of energy can be either absorbed or emitted.

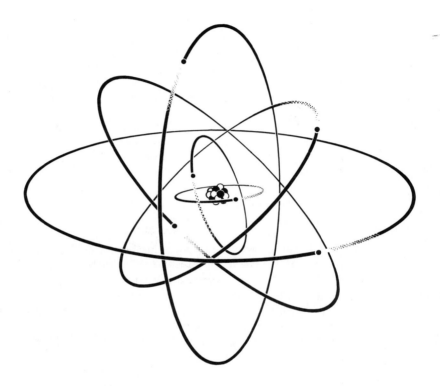

Figure 2-8.   The Nuclear Atom
Rutherford proposed that atoms consist of dense, positively charged nuclei orbited by tiny, negatively charged electrons. Each atom is therefore like a miniature solar system. Today we know that the nucleus itself is composed primarily of two types of particles: protons and neutrons.

Imagine an experiment in which you shine a continuous spectrum of white light onto some gas or vapor. Since the white light contains photons of all wavelengths, and since the energy of a photon is determined by its wavelength, you are literally bombarding the gas with photons of all energies. Some photons will have exactly the right amount of energy to boost electrons from low orbits to high orbits in the atoms in the gas. These atoms will therefore selectively absorb these particular photons, letting all the others pass by. By absorbing photons of a particular energy, atoms in the gas will have extracted light of a particular wavelength from the original beam. This is the true meaning of absorption lines. Each line corresponds to one particular electron transition between orbits of the atoms of a particular element. Spectral lines tell us the intimate details of the quantum mechanical structure of atoms.

When an atom has absorbed photons so that electrons are orbiting the nucleus in unusually high orbits, the atom is said to be *excited*. As electrons in an excited atom drop back down to their usual orbits, the atoms give up photons of precisely the same total amount of energy they originally absorbed. Of course, the atom does not know left from right, and therefore it emits photons in random directions. This explains Kirchhoff's laws. In looking at a continuous spectrum through a gas, we see absorption lines because the atoms have absorbed photons of specific wavelengths. At oblique angles we see only emission lines as the excited atoms give up their photons and return to their original states.

During the decades following Fraunhofer's discovery of hundreds of absorption lines in the sun's spectrum, astronomers were fascinated by the prospects of studying the spectra of stars. Unfortunately, the astronomers really did not know what they were doing. After all, no one really understood exactly why spectral lines existed in the first place. Nevertheless, astronomers were amazed to find that the spectra of stars vary widely from one to the next. For example, Figure 2-10 shows portions of spectra of three stars. As you can see, they are strikingly different.

For lack of anything better to do, astronomers collected the spectra of stars and grouped them according to the intensity of different spectral lines. Each group in this classification scheme was

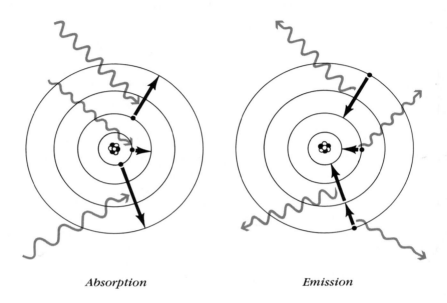

Absorption   Emission

Figure 2-9. Absorption and Emission

Atoms absorb photons when electrons jump from low orbits to high orbits. As the electrons drop back down, the atoms emit photons. Atoms can absorb or emit photons only if the energy of these photons corresponds exactly to the energy gained or released in a particular electron transition between orbits.

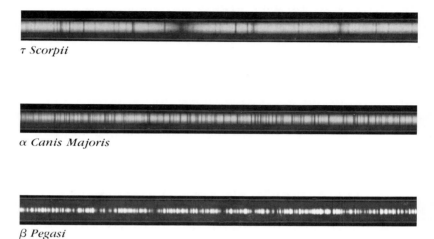

τ Scorpii

α Canis Majoris

β Pegasi

Figure 2-10.   Some Stellar Spectra
The violet portions (roughly 4,000 to 4,200 Å) of the spectra of three stars are shown. Notice how vastly different the spectra appear even though all three stars have essentially the same chemical composition. The upper spectrum is of a hot star (τ Scorpii) whose surface temperature is 25,000°K. Procyon (α Canis Majoris) has a surface temperature of 7,000°K. The lower spectrum is of a cool star (β Pegasi) whose surface temperature is 3,000°K. (Hale Observatories.)

assigned a letter of the alphabet. For example, the spectra of "A stars" show strong hydrogen lines. "B stars" show weaker hydrogen lines, but instead have noticeable helium lines. The spectra of "K stars" are dominated by the spectral lines of metals, whereas "M stars" exhibit broad bands of spectral lines because of the presence of molecules such as titanium oxide.

No one was about to believe that some stars are made of hydrogen while others have nothing but metals or titanium oxide. Very much to the contrary, we today know that *all* stars have almost the *same* chemical composition. Hydrogen, the lightest element, is by far the most abundant. Hydrogen alone constitutes from 60 to 80 percent of the mass of most stars. Helium, the second lightest element, happens to be the second most abundant. Hydrogen and helium together account for 96 to 99 percent of the mass of a star. That leaves less than 4 percent for all the other elements combined. But if the chemical compositions of all stars are essentially the same, why are their spectra so vastly different? How can it be that a star containing mostly hydrogen shows only absorption lines of metals in its spectrum? The answers finally came from the ideas of Planck, Kirchhoff, and Bohr.

In order for the spectral lines of a particular chemical to appear in a star's spectrum, a few very specific requirements must be met. First of all, the chemical must be present in the star's atmosphere. That's obvious. But in addition, the star's surface temperature plays an equally important role. If the star is very hot, then atoms in the star's atmosphere are bombarded with lots of high-energy, short-wavelength photons. Instead of raising electrons from low orbits to high orbits, these photons may simply tear the electrons completely off the atoms. Such atoms are said to be *ionized*. Alternatively, if the star is very cool, then the atoms in the star's atmosphere are bathed in low-energy, long-wavelength photons. These photons may not have enough energy to get the electrons up out of their low-lying orbits. In either case—too hot or too cool—the atoms do not produce spectral lines. Only if the temperature is "just right" can the atoms effectively absorb photons and produce spectral lines.

Consider hydrogen, for example. Only if the star's surface temperature lies in the range of 7,500 to 11,000°K does the star's spectrum exhibit strong hydrogen lines. A blackbody in this temperature range emits many photons with just the right energy to boost electrons in hydrogen atoms from low to high orbits. In hotter stars, like Rigel and Spica, the enormous number of ultraviolet photons easily ionize hydrogen atoms. In cooler stars, like the sun, there are not enough short wavelength photons to excite an appreciable number of hydrogen atoms. Thus, even though hydrogen is very abundant in all stars, hydrogen lines show up only if the star's surface has the right temperature.

Similar arguments apply for all chemicals. In the hot stars virtually all atoms are ionized except for helium. Helium atoms manage to hold onto their electrons quite well and thus stars with surface temperatures around 20,000°K (for example, Rigel and Spica) have helium lines in their spectra.

At about 10,000°K (for example, Sirius, and Vega), the star is too cool for helium lines and all the metals are still ionized. But this surface temperature is just right for hydrogen lines.

At roughly 7,000°K (for example, Canopus and Procyon) some metals such as iron and chromium manage to retain all their electrons. And for stars whose temperature is 5,000°K (for example, the sun and Capella) the spectral lines of many metals are seen.

On cooler stars (for example, Arcturus and Aldebaran around 4,000°K) some atoms can combine to form molecules. In fact, on the coolest stars (for example, Betelgeuse and Antares at 3,000°K), the temperature is so low that titanium atoms easily stick to oxygen atoms, producing titanium oxide. Even though titanium is a very rare element, the structure of the titanium oxide molecule is such that it efficiently absorbs vast quantities of photons at visible wavelengths.

Stellar temperature is the key to understanding stellar spectra. Almost exactly a full century after Fraunhofer first saw absorption lines in the sun's spectrum, science had progressed to the point that we could begin to understand the true meaning of starlight.

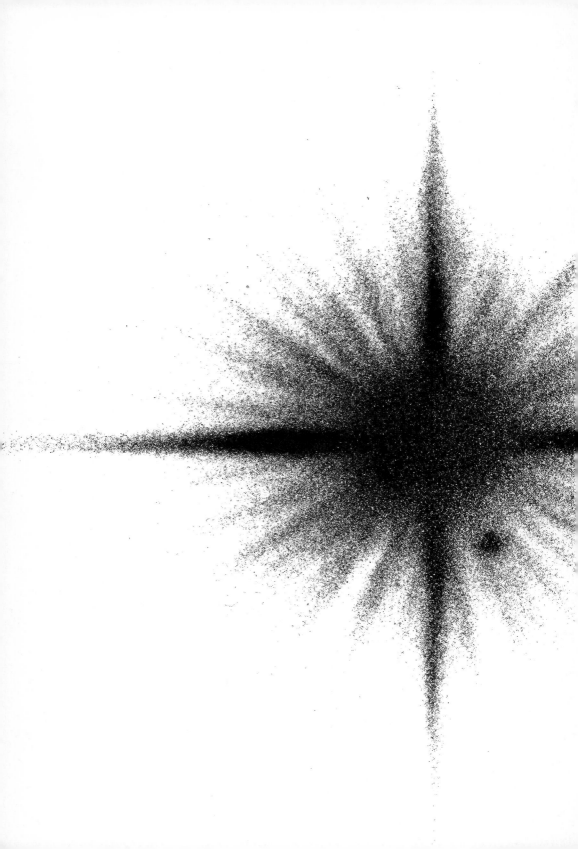

3

The Meaning of Starlight

In 1823, almost a decade after he had first examined the sun's spectrum, Joseph Fraunhofer discovered that stars exhibit absorption lines in their spectra. Following up on the work of Fraunhofer, and armed with the discoveries of Bunsen and Kirchhoff, in 1864 Sir William Huggins identified some lines in stellar spectra that were caused by chemical elements that exist on earth. This conclusively proved that we can discover what the stars are made of.

While Huggins was beginning to discover the chemical composition of stars in England, Angelo Secchi was attempting spectral classification in Italy. Secchi's scheme of classifying stars into groups according to the arrangement of lines in their spectra was modified and enlarged over the years. But a real understanding of stellar spectra did not come until the beginning of the twentieth century.

In order to appreciate the true significance of stellar spectra, it was necessary to know that light consists of photons and that the energy carried by a photon depends on its wavelength. Astronomers had to learn that electrons orbit the nuclei of atoms and that transitions between allowed orbits produce spectral lines. Only then could they begin to fully appreciate the fact that *the spectrum of a star tells us the temperature of the star's surface.*

There is another, very different way of measuring the temperatures of stars. If you look carefully at the stars in the sky, you will notice that they are not all exactly the same color. Some stars are a little bluish, whereas others appear slightly reddish. But from Wien's law and the work of Max Planck, we know that these color differences reflect differences in temperature. For example, Figure 3-2 shows the blackbody curves for three temperatures. These curves approximate the actual distribution of energy emitted by stars. For a star at 25,000°K, the curve peaks at ultraviolet wavelengths. Because the curve is so strongly skewed toward the ultraviolet, much more blue light is emitted than red light. This star therefore looks blue. In contrast, for a cool star at 3,000°K, the curve peaks in the infrared. Since more red light is given off than blue, the star looks reddish. And finally, for a star at about 6,000°K, the curve peaks in the middle of the visible spectrum. Since all wavelengths are emitted in roughly equal amounts, this star has a yellowish-white appearance.

Of course, stars are not perfect blackbodies, but the distribution of energy that is emitted from a star's surface is usually close to a

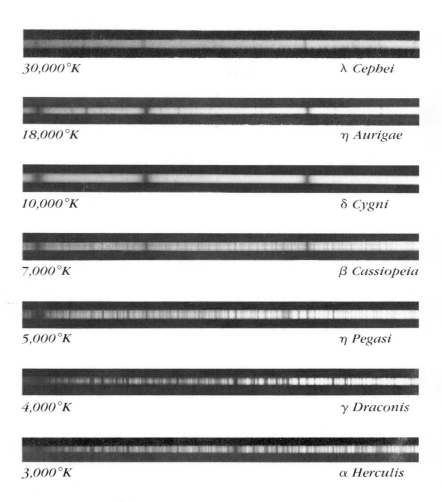

30,000°K   λ Cephei

18,000°K   η Aurigae

10,000°K   δ Cygni

7,000°K   β Cassiopeia

5,000°K   η Pegasi

4,000°K   γ Draconis

3,000°K   α Herculis

Figure 3-1. Stellar Spectra
These seven stellar spectra cover the full range of stellar temperatures. The appearance of a star's spectrum is directly determined by its surface temperature. The spectra of the hottest stars are dominated by helium lines since all other elements are highly ionized. In the coolest stars, the temperature is so low that atoms can combine to form molecules. Thus their spectra are dominated by bands of molecular lines. (Hale Observatories.)

49

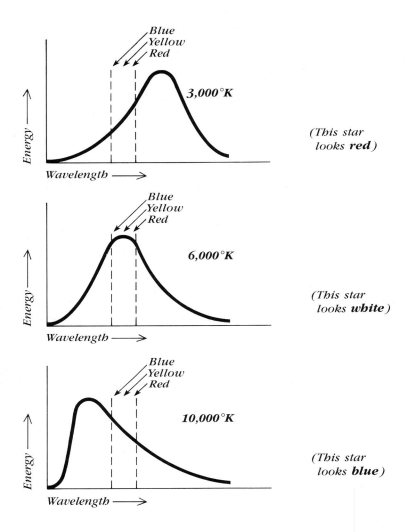

Figure 3-2. Temperatures and Colors

Blackbody curves for three temperatures are shown. Notice that the relative amounts of light emitted in various colors are directly related to the temperature. Since the actual distribution of energy emitted by a star closely approximates a blackbody curve, astronomers can determine the surface temperature of a star by accurately measuring the colors of the light it radiates.

blackbody curve (spectral lines cause small wiggles and dips, as shown in Figure 2-1). From the work of Max Planck, we have precise mathematical formulas for blackbody curves. For a particular temperature, we know exactly the relative intensities of light emitted at various wavelengths. Conversely, by measuring the relative intensities of light from a star at various wavelengths, we should be able to deduce the star's temperature.

This is the central idea behind a technique called *photoelectric photometry.* Using colored filters and a light-sensitive photoelectric cell mounted at the focus of a telescope, astronomers can very accurately measure the intensity of light from a star in several wavelength ranges. By comparing the relative intensities of light in various colors, they can easily deduce the star's surface temperature. This is often a lot easier than taking spectra because, instead of waiting to develop a spectroscopic plate, an astronomer gets the data immediately right at the telescope.

By either stellar spectroscopy or photoelectric photometry, astronomers discover the surface temperatures of stars. These observations tell us that stellar temperatures lie in the range of 3,000 to 30,000°K. Like the sun at 5,800°K, the majority of stars have surface temperatures near the cool end of the range. But there are a few very hot stars in the sky such as Rigel, β Crucis, β Centauri, Spica, and Achernar, all of which shine with a brilliant blue light.

Around the turn of the twentieth century, several teams of American astronomers participated in extensive and laborious observing programs. The resulting data would soon provide some important clues to the mysteries of the stars. For example, at Yerkes Observatory, Frank Schlesinger used photography for the measurement of stellar parallaxes. As I said in Chapter 1, parallax is the key to determining the distances to stars. And once the distance to a star is known, its luminosity can easily be computed. Thus, the work of people like Schlesinger began to give us knowledge of the true brightnesses of many stars across the sky.

While Schlesinger was starting his parallax survey at Yerkes, work was already progressing at full steam on a spectroscopic survey at Harvard. In particular, Annie J. Cannon (who single-handedly examined the spectra of nearly 400,000 stars!) and her associates discovered that the spectra of stars could be arranged in a particular order so that the intensity of spectral lines varied smoothly from one to the next. This ordering of stellar spectra is shown in Figure 3-1. Of course, no one at that time fully realized that this amounted to an ordering according to the surface temperature of stars.

At about the same time that Niels Bohr was beginning to discover the quantum mechanical structure of atoms, two astronomers invented an ingenious way of looking at all this new data about the brightnesses and spectra of stars. Between 1911 and 1913, independent of each other, Ejnar Hertzsprung in Denmark and Henry Norris Russell in the United States hit upon the idea of making a graph on which the absolute magnitudes of stars could be plotted against Annie Cannon's ordering of spectra. This simple, straightforward plotting of data is called a *Hertzsprung-Russell diagram.* This graph was soon destined to become the single most important diagram in all of modern astronomy.

Any graph in which you plot the brightness (*or* luminosity *or* absolute magnitude) of stars against their spectral type (*or* surface temperature *or* color) is a Hertzsprung-Russell diagram. Astronomers who spend most of their time at telescopes usually think of plotting absolute magnitudes against spectral type. Theoretically inclined astronomers, who prefer the air-conditioned comfort of a computer room over the sometimes bitter cold of an isolated mountain top, usually think of plotting luminosity against temperature. Either way, it is essentially the same graph.

A Hertzsprung-Russell diagram is shown in Figure 3-3. Every dot represents a star whose luminosity and surface temperature have been measured. The remarkable fact is that the data are not scattered randomly all over the graph. Instead the dots are grouped together in three distinct regions.

The first striking feature we notice about the Hertzsprung-Russell diagram is that a large number of the dots, each representing a star, fall along a line extending from the upper left to lower right corners of the graph. The upper left-hand corner of the H-R diagram is for hot, bright blue stars while the lower right-hand corner is reserved for cool, dim red stars. The band of dots connecting these two regions is called the *main sequence.* Any star whose dot falls along the main sequence is called a *main sequence star.*

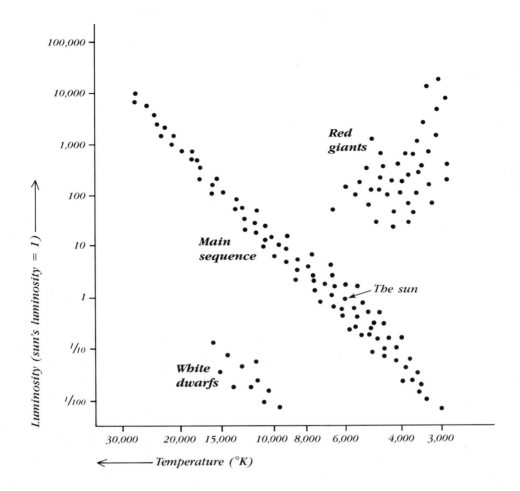

Figure 3-3. **The Hertzsprung-Russell Diagram**
A Hertzsprung-Russell diagram is simply a graph on which the luminosities of stars are plotted against their surface temperatures. Every dot represents a star whose brightness and temperature have been measured. The fact that the data fall into three distinct regions means that there are three very different kinds of stars in the sky.

Almost every star you see in the sky is a main sequence star. The sun's luminosity and surface temperature are just right to place its dot squarely on the main sequence. The sun is therefore a typical main sequence star.

The existence of the main sequence on the H-R diagram means that, for these stars, there is a correlation between surface temperature and brightness. And that is what we expected from Stefan's law: The hotter an object is, the brighter it glows. Cool main sequence stars are dim, whereas hotter main sequence stars shine more brilliantly.

Set apart from the main sequence, near the upper right-hand corner of the H-R diagram, is a second prominent grouping of dots. These dots represent stars that are both cool and bright. But how can a low temperature star shine so brilliantly? Is this a violation of Stefan's law?

Stefan's law tells us that each square mile on the surface of a cool star emits a lot less light than a corresponding square mile on the surface of a hot star. That's the way things work. But a cool star *could* produce a lot of light *if* it had a lot of surface area. That is what is indicated in the upper right-hand corner of the H-R diagram. Stars whose dots appear in that portion of the H-R diagram are truly enormous. Although each square mile of surface area produces only a moderate amount of light, the combined influence of trillions upon trillions of square miles results in a dazzlingly bright star. Since the surface temperatures are low, these stars emit mostly reddish light (that's Wien's law). Since the surface areas are so extensive, these stars must be gigantic. They are *red giants*.

Almost every reddish-appearing star you can see in the sky is a red giant. Betelgeuse, Antares, Aldebaran, and Arcturus are well-known examples. They are thousands of times larger than their counterparts on the main sequence. Red giants are so big that if one were located at the center of the solar system, the star's surface would lie between the orbits of Earth and Mars. They are the largest stars in the universe.

In sharp contrast to the red giants, there is a third and final significant grouping of dots on the Hertzsprung-Russell diagram. A few dots sprinkled in the lower left-hand corner represent stars that are both dim and hot. Just as a cool star could be bright if it has a large surface area, a hot star could be dim if it has a small surface area. Indeed, these hot, dim stars are small. They are typically about the same size as the earth. And in view of the blue-white light emitted from their hot tiny surfaces they are called *white dwarfs*.

The first great lesson to come from the Hertzsprung-Russell diagram is that there are three *very* different kinds of stars in the sky. Main sequence stars, red giants, and white dwarfs are about as different from each other as apples, watermelons, and raisins.

But why are there vastly different kinds of stars? What actually happens that makes red giants so huge and white dwarfs so small? Are these different kinds of stars related in any way? Indeed, why are the dots on the H-R diagram grouped in various regions in the first place?

Answers to all these questions provide the greatest lesson to come from the Hertzsprung-Russell diagram. The appearance of the H-R diagram, the arrangement of the dots, and how the various kinds of stars are related intimately involves the life cycles of the stars themselves. If you know how stars are born, what happens to them as they grow old, and where they go after they die, then you will have finally discovered the inner meaning of the Hertzsprung-Russell diagram.

In order to understand the life cycles of the stars in the sky, the astrophysicist cannot simply sit down and dream up a story about what might have happened. Instead, he must construct theoretical models of stars using the laws of physics and calculate how his model stars change with age. Of course he must constantly check the results of his calculations against the astronomer's observations. Only then can the astrophysicist be sure that he is on the right track.

To even begin to understand stellar evolution, the astrophysicist must first know exactly what a star is. Spectroscopy tells us how hot stars are and what they are made of. We know that a star's parallax permits us to calculate the star's true brightness. We have even learned that the laws of physics can be used to present arguments about a star's size. But in addition to all this vital information, we also need to know how much matter there is in a star. Whether in tons or kilograms, we need to know the *masses* of the stars.

This would be a hopeless task if it were not for the fact that there are many *binary* or *double stars* in the sky. Indeed, about half of the stars you see in the sky are not single stars like the sun. Instead they consist of two stars in orbit about each other just as the earth and the moon revolve about their common center. From observing binary stars, astronomers gain crucial information about the masses of stars.

The first double star was discovered in 1650 by the Italian astronomer Jean Baptiste Riccioli. This star, Mizar, can be easily seen in the "handle" of the Big Dipper. In 1656, Christian Huygens discovered that θ Orionis, the middle star in the "sword" of Orion, actually consists of three stars. And in 1664, Robert Hooke found that γ Arietis is really a pair of stars.

During the next century, astronomers would continue to stumble across binary stars during the course of their observations. Remarkably, everyone seemed to think that these double stars were nothing more than the result of a chance alignment between two stars in the sky. In other words, each of these binaries was assumed to consist of a nearby star and a distant star that just happen to be located in almost exactly the same direction. It was not until 1779 that the astronomer Christian Mayer seriously speculated on "the possibility of small suns revolving around larger ones."

Realizing the tremendous potential importance of double stars, the German-born English astronomer Sir William Herschel set about the business of systematically searching for double stars in the sky. In 1782, after only two years of observation, Herschel had observed and catalogued 269 double stars "227 of which, to my present knowledge, have not been noticed by any person." Within the next two years, he had found 434 more.

Over the next few years, Herschel became interested in other topics. But in 1797, he went back to observing double stars. In examining some of the double stars he had discovered twenty years earlier, Herschel was amazed to find that many of them had shifted their relative positions. This shifting of positions (as shown, for example, in Figure 3-4) proved to Herschel that these double stars "are not merely double in appearance, but must be ... real binary combinations of two stars, intimately held together by the bonds of mutual attraction." Double star astronomy was off and running.

Over the next century, many astronomers devoted much of their lives to observing binary stars. Typical results of many years of observing one particular double star are shown in Figure 3-5. The

56

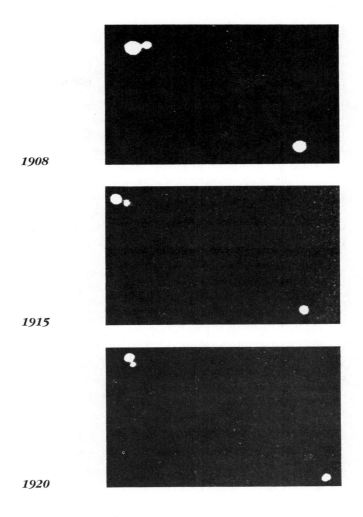

1908

1915

1920

Figure 3-4. The Double Star Kruger 60
Careful observations over many years often reveal that two stars located near each other in the sky are actually in orbit about each other. Indeed, about half of the stars in the sky are double stars. Kruger 60, in the constellation of Cepheus, is an excellent example. (Yerkes Observatory.)

important point is that once we have observed the orbits of the stars in a binary, we can make calculations about the masses of these stars. It is the exact opposite of the problem of sending astronauts to the moon. In the Apollo missions, we know the masses of the earth and the moon and we want to calculate an orbit for the astronauts. In a binary star we see the orbit and want to calculate the masses of the stars.

Well, often things do not work out very easily for the astronomer. Many times the best he can do is arrive at a reasonable range for the mass of a star in a binary ("It's more than thus-and-such, but less than so-and-so"). Or perhaps the best he can do is arrive at a combined mass of both stars together. In many binary stars, the two stars are so close that it is impossible to view or resolve them individually, even with the most powerful telescopes. Yet the astronomer is sure that a particular star that appears single is really double because of certain phenomena associated with it. For example, in observing a single-appearing star, an astronomer might notice that the brightness of the star seems to vary in a periodic fashion, as shown in Figure 3-6. The obvious interpretation of this *light curve* is that the star is really double and that the two stars alternately pass in front of each other. This is called an *eclipsing binary*. From the shape of the light curve the astronomer can deduce important information about the masses of the stars.

Another type of unresolved binary (that is, one in which the two stars cannot be seen individually) is a *spectroscopic binary*. Occasionally an astronomer finds that the spectrum of a single-appearing star shows *two* sets of spectral lines. The obvious conclusion is that the single-appearing star is really a double star. In addition, the two sets of spectral lines are observed to shift back and forth as the two stars in their orbits alternately approach or recede from the earth. By measuring the shifting of the spectral lines, the astronomer can gain significant information about the masses of the stars.

The final result of almost two centuries of observing binary stars is that we have a good idea about the masses of stars. Astronomers strongly prefer to express the masses of the stars in terms of the mass of the sun: in *solar masses*. This is a lot more convenient than speaking of trillions of tons or quadrillions of kilograms.

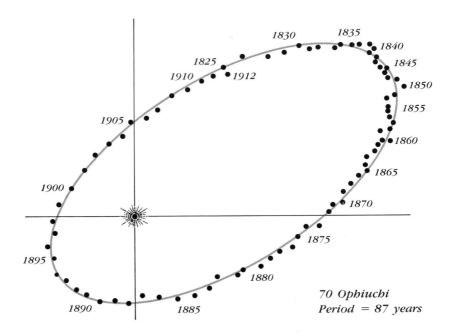

Figure 3-5. The Orbit of 70 Ophiuchi
By plotting the observations of a binary star over the years, it is possible to draw the orbit of one star about the other. Once the orbit is known, calculations concerning the masses of the stars can be made. This is the orbit of a faint double star in the constellation of Ophiuchus.

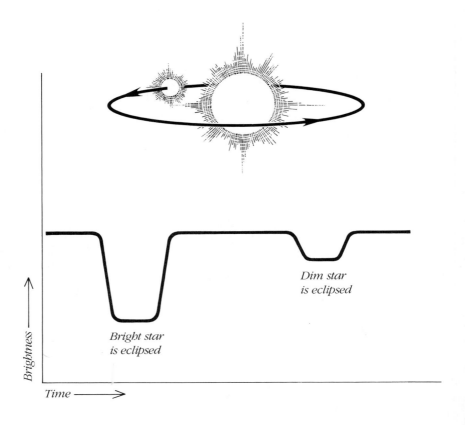

Figure 3-6.   An Eclipsing Binary
An eclipsing binary is a double star oriented so that the two stars alternately pass in front of each other. Usually astronomers discover eclipsing binaries by noting that the brightness of a single-appearing star exhibits periodic, temporary depressions. From the shape of the resulting light curve, many details about the binary star can be deduced.

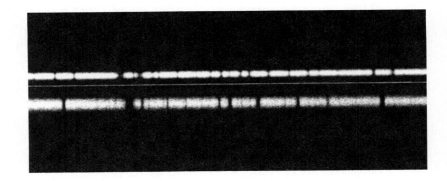

Figure 3-7. A Spectroscopic Binary
Usually astronomers discover spectroscopic binaries by noting that the spectrum of a single-appearing star exhibits two sets of spectral lines. From observing how these sets of spectral lines shift as the two stars revolve about each other, many details about the binary system can be deduced. These are two spectra of the spectroscopic binary κ Arietis. (Lick Observatory Photograph.)

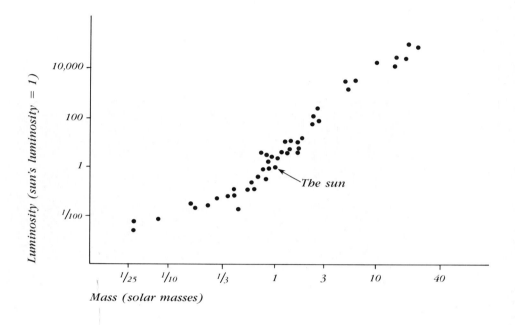

Figure 3-8. The Mass-Luminosity Relation
The masses of main sequence stars are directly correlated with their luminosities. Dim stars have low masses, whereas bright stars have high masses. This relationship holds only for main sequence stars and does not apply to red giants or white dwarfs.

Obviously, the mass of the sun is exactly one solar mass (nearly two billion billion billion tons). The masses of other stars in the sky fall in the range of one-tenth solar mass up to about 50 solar masses. High-mass stars (that is, stars containing 40 to 50 solar masses of matter) are extremely rare, however. Most stars in the sky contain about one solar mass or less.

As reliable measurements of stellar masses became available, an interesting trend started to emerge. For main sequence stars, the mass of a star was directly correlated with its luminosity. Low-mass stars were dim while high-mass stars were bright. This is called the *mass-luminosity relation* and it is diagrammed in Figure 3-8. This relationship holds only for main sequence stars and does not apply to red giants or white dwarfs.

At this point we see that the main sequence on the Hertzsprung-Russell diagram is a continuous arrangement of stars according to surface temperature, luminosity, *and mass*. The dim, cool, red stars whose dots appear in the lower right-hand corner of the H-R diagram are very low-mass stars. The bright, hot, blue stars whose dots appear in the upper left-hand corner of the H-R diagram are very high mass stars. Stars with intermediate luminosities and temperatures have intermediate masses.

We now have all the necessary information and data to develop a coherent, complete theory of stellar evolution. But first, perhaps we should pause to examine one particular well-known main sequence star. This particular star is typical of the average main sequence star in virtually every respect. It is of special interest to the astronomer simply because it happens to be the nearest star in the sky. At only 93 million miles from Earth, it is the only star whose surface can be examined in great detail.

4

Our Sun –
A Typical Star

In many respects, the sun is the most important star in the sky. Radiation from the sun provides the heat and light without which the earth would be a barren, frigid wasteland. The sun bestows the energy that powers the currents in the oceans and the winds in the air. The sun dictates what portions of the earth shall be arid deserts, lush rain forests, or frozen tundra. The sun dramatically influences our behavior and controls major portions of our daily lives — whether you wear a heavy fur coat or a swimsuit, whether you are awake or asleep, is largely determined by the position of the sun in the sky relative to your location on our planet.

In many respects, the sun is the least important star in the sky. Nothing of any significance distinguishes the sun from billions upon billions of other stars in our Milky Way galaxy — it is, in every sense, an ordinary garden-variety star. It has a typical size, a typical mass, a typical temperature, and a typical luminosity. Indeed, the only reason our star seems different from any other star is that we just happen to be going around it.

The first person to take a good look at the sun was Galileo Galilei in 1610. Galileo did not invent the telescope; he was just the first person to point it up towards the sky. In doing so, he saw a few, small dark spots on the sun's surface. These temporary features are called *sunspots*.

By observing the same group of sunspots from one day to the next, Galileo soon concluded that the sun rotates. He found that the sun's period of rotation is slightly less than one month.

Actually, the sun's rotation is more complicated than Galileo would have supposed. The sun does *not* rotate like a rigid body. Instead, as first discovered by Richard Carrington in 1859, the equatorial regions rotate faster than the polar regions. A sunspot near the solar equator takes only 25 days to go around the sun once. At 30° north or south of the equator, a sunspot takes 27½ days to complete one rotation. At 75° north or south of the equator, the rotation period is about 33 days, and at the poles it may be as long as 35 days. This phenomenon is called *differential rotation* because different parts of the sun rotate at slightly different speeds.

Careful observations of the sun over many years reveal that the numbers of sunspots change in a periodic fashion. In some years there are many spots, but in other years there are almost none. This

Figure 4-1.   A Galaxy
*A galaxy is a huge collection of billions upon billions of stars. This
galaxy, in the constellation of Andromeda, happens to have an
overall shape and size quite similar to our own. Our galaxy is
100,000 light years in diameter and it contains about 200 billion
stars. Our star, the sun, is located nearly 30,000 light years from the
center of our galaxy. (Lick Observatory Photograph.)*

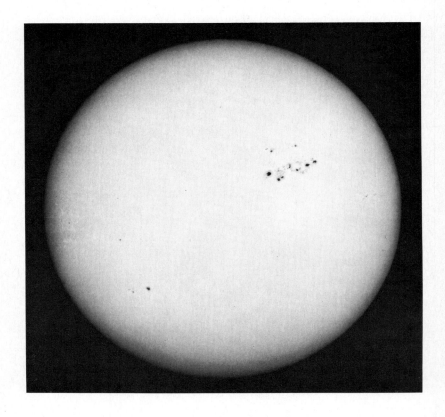

Figure 4-2. The Sun
The sun is by far the largest and most massive object in the solar system. The sun's diameter (1 million miles) is 100 times larger than the diameter of the earth. The sun's mass is 333,000 times greater than the mass of the earth. A telescopic view of the sun's surface often reveals sunspots in the solar photosphere. (Kaufmann Industries, Inc.)

phenomenon, first noticed by Heinrich Schwabe in 1851, is called the *sunspot cycle.* As shown in Figure 4-4, *sunspot maxima* occurred in 1948, 1959, and 1970. *Sunspot minima* occurred in 1954, 1965, and 1976.

The differential rotation of the sun is responsible for a wide range of phenomena in the solar atmosphere that are associated with the sunspot cycle. To understand why this is so, we first must realize that the sun, like the earth, has a natural magnetic field. Indeed, the strength of the sun's magnetic field is roughly of the same intensity as the earth's. But because of the sun's differential rotation, the magnetic field inside the sun's surface becomes wrapped around the sun. After all, the United States would become twisted if Mexico went around the earth faster than Canada! (Fortunately, the earth is rigid and does not exhibit differential rotation; the day is 24 hours long in both Mexico City and Quebec.)

After many solar rotations, the sun's magnetic field has become so stretched and twisted that the magnetic field occasionally buckles and breaks through the solar surface. In places where the twisted magnetic field ruptures the sun's surface, the boiling and bubbling of the hot gases are severely inhibited. Because the boiling and bubbling are reduced, these locations will have a lower temperature than the surrounding solar surface. But if the temperature goes down, then (according to Stefan's law) the amount of light emitted from these regions must also decrease. These regions therefore appear significantly darker than the surrounding solar surface: they are the sunspots we see.

This theory of sunspot formation was first proposed by Horace W. Babcock in 1961. The theory takes support from the fact that sunspots are cool regions on the sun. The temperature in the middle of a sunspot is typically 4,600°K, compared to 5,800°K for the surrounding normal solar surface. In addition, ever since the pioneering work of George E. Hale in 1908, it has been known that sunspots contain intense magnetic fields. The strength of the magnetic field inside a sunspot is typically hundreds or thousands of times stronger than the sun's overall magnetic field.

Babcock's theory explains that it takes 22 years for a complete cycling of the sun's magnetic field through the sun's atmosphere.

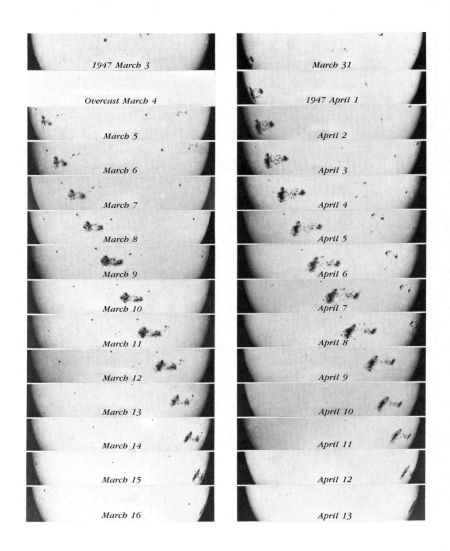

Figure 4-3. The Sun's Rotation

By observing the same group of sunspots from one day to the next, Galileo found that the sun rotates once in about four weeks. Actually, the equatorial regions of the sun rotate somewhat faster than the polar regions. This series of photographs shows the same large sunspot group during 1 ½ solar rotations. (Hale Observatories.)

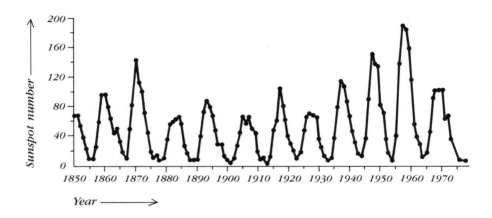

Figure 4-4.   The Sunspot Cycle
The number of sunspots on the sun varies in a regular and periodic fashion. Large numbers of sunspots were seen in 1948, 1959, and 1970. The sun was almost completely devoid of sunspots in 1954, 1965, and 1976. The next sunspot maximum is due to occur in 1981.

71

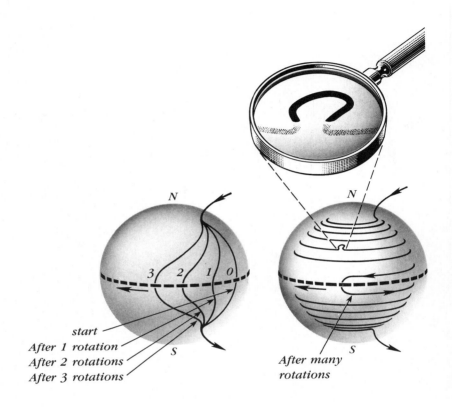

Figure 4-5. *Differential Rotation and Sunspot Formation*
Due to the sun's differential rotation, lines of magnetic field get wrapped around the sun. After many rotations, the magnetic field is so stretched and wrapped up that it buckles. Sunspots form where the buckled magnetic field has broken through the sun's surface.

Sunspot maxima occur when the magnetic field is most severely twisted just below the solar surface. Sunspot minima occur when the magnetic field is straightened out in the sun's upper atmosphere. Two wrapping and unwrapping sequences occur during each 22-year cycle.

Sunspots are just one of many phenomena that occur as twisted magnetic fields puncture the solar surface. With solar telescopes, special filters, and spectroheliographs, astronomers study solar flares, plages, faculae, and filaments that often occur near sunspots. Nevertheless, a sunspot just happens to be the easiest thing to see through a telescope in ordinary white light.

When you look through a telescope in ordinary white light, you see a region in the solar atmosphere called the *photosphere* (literally "sphere of light"). When astronomers speak of the sun's "surface," they usually mean the photosphere. The sun really does not have a surface like the earth or the moon. The photosphere is simply the layer in the solar atmosphere from which an overwhelming percentage of visible light comes. Figure 4-2 and 4-6 are photographs of the photosphere.

Extending to an altitude of 6,000 miles above the photosphere is a second layer in the solar atmosphere called the *chromosphere* (literally "sphere of color"). Using a red filter that transmits light only at a wavelength centered on one particular spectral line of hydrogen (that is, at 6,563 Å centered on the Hα line, thereby excluding the blinding white light of the photosphere), astronomers can study phenomena in the chromosphere. Jetlike spikes of gas called *spicules* are often seen rising upward through the chromosphere. These spicules (see Figure 4-7) shoot upward at speeds of roughly 20 miles per second and play an important role in transporting energy from the photosphere to the thin, outer layers of the sun's atmosphere.

Extending for millions of miles above the chromosphere is the third and final layer in the sun's atmosphere. This region, called the *corona,* can be seen most easily during a total solar eclipse. During the few precious moments when the moon blocks out the blinding solar disc, a delicate lacework of glowing gases is seen all around the sun. The innermost portions of the solar corona are shown in Figure 4-8.

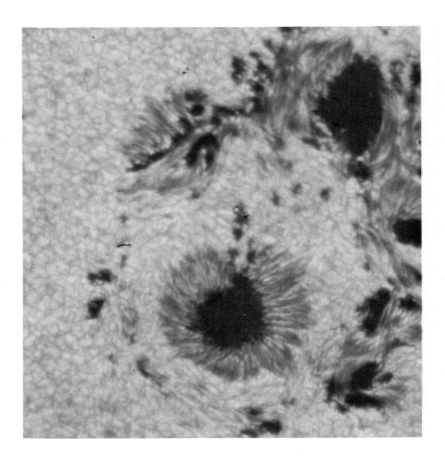

Figure 4-6. The Photosphere
*This extraordinary photograph of the solar surface was taken
from a balloon. Notice the granular structure of the photosphere.
Each granule is a column of hot, rising gases—literally the
boiling and bubbling of the solar surface. Each is about 500 to
1,000 miles in diameter and lasts for about 5 to 10 minutes.
(Project Stratoscope; Princeton University.)*

Figure 4-7. Spicules in the Chromosphere
Jets of hot gas, called spicules, are visible in this Hα photograph of the sun's limb. An individual spicule may last for 10 minutes as it surges upward transporting matter and energy from the photosphere to the corona. (Hale Observatories.)

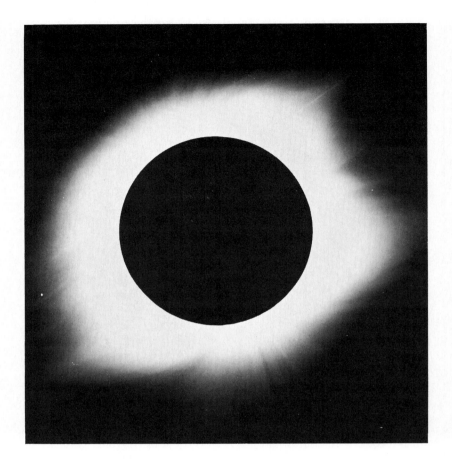

Figure 4-8. The Corona
The thin, outermost portions of the solar atmosphere can be seen during a solar eclipse. The solar corona extends millions of miles out into space and eventually merges with the solar wind. Only the inner portions of the corona are visible in this photograph, which was taken during an eclipse on September 22, 1911. (Lick Observatory Photograph.)

While observing the sun's limb, astronomers often see huge eruptions of gas extending far out into the corona. These eruptions, called *prominences,* rise tens or hundreds of thousands of miles above the solar surface. Unlike spicules, which last for a few minutes, a prominence may hang for weeks above the sun. Prominences, such as the one shown in Figure 4-9, are usually associated with sunspots. As the magnetic field erupts through the photosphere to form a sunspot, gases in the hot corona are cooled. As the gases cool, they become visible, especially when viewed through filters that block out the blinding white light from the photosphere.

In 1940, the Danish astronomer Bengt Edlén made an important discovery about the spectrum of the sun's corona. For many years, scientists had been mystified by unfamiliar spectral lines detected in the corona. Edlén showed that these mysterious lines were caused by highly ionized atoms such as silicon and iron. In other words, for example, a particular set of spectral lines might be produced by iron atoms that had been stripped of 16 of their normal 56 electrons.

Atoms lose so many electrons only if the temperature is extremely high. At high temperatures, atoms can be stripped of many of their electrons by high-speed collisions either with other atoms or with highly energetic, short-wavelength photons. Because, as Edlén proved, the atoms in the corona have lost many of their electrons, the corona must be very hot. Indeed, today we realize that the temperature in the corona is about 2 million degrees Kelvin.

In view of these discoveries about the corona, we are faced with the remarkable fact that it gets hotter as you move higher in the solar atmosphere. In the photosphere, the temperature is about 6,000°K. Moving upward into the chromosphere, we find that the temperature rises to about 20,000°K. And thousands of miles up in the corona, the temperature climbs to 2,000,000°K.

It is believed that the corona's high temperature is caused by turbulence down in the photosphere. Bubbling and boiling gases in the solar surface generate an enormous amount of noise, technically called *acoustic energy.* As the acoustic energy in the form of shock waves crashes upward through the solar atmosphere, the gases in the corona are heated to extremely high temperatures. In other words,

77

the corona is so very hot because the violently boiling photosphere is so very noisy.

The sun is important to astronomers simply because it is the only star that we can study at close range. And indeed the solar atmosphere is filled with fascinating phenomena that could easily occupy a lifetime of study. Nevertheless, one of the greatest lessons to come from the sun concerns processes that we cannot see with telescopes: the thermonuclear processes in the sun's center.

At the turn of the twentieth century, scientists realized that they were in an embarrassing position. They simply could not explain why the sun shines. The source of this embarrassment was geology rather than astronomy. Since the late 1800s, geologists had been discovering extremely ancient rocks. These rocks strongly suggested that the earth was very old—perhaps a few billion years old. And yet, scientists could not think of any processes by which the sun could shine for more than a hundred million years.

A piece of coal the size of the sun would burn for only a few thousand years. Obviously chemical combustion was not the source of the sun's energy.

The two great nineteenth century physicists H. von Helmholtz and Lord Kelvin had shown that a gradual contraction of the sun could provide energy for a total of 100 million years. They said that as the sun gradually shrank in size, the compressed gases were heated. The hot, glowing gases thereby produced the sun's light. That was the accepted explanation until geologists started digging up extremely ancient rocks—rocks that had been formed earlier than the sun was thought to have existed. And it seemed absurd to think that the earth formed billions of years ago and then only a few million years ago the sun just happened to come along.

Some astronomers argued vigorously that the geologists were wrong about the age of the rocks they had dug up. But the geologists did not seem to mind; they just kept on digging up more ancient rocks. The final result was that classical physics was plunged into a major crisis over the most childlike of questions: "Why does the sun shine?"

At about the same time that astronomers were realizing the extent of their lack of knowledge, some important developments were occurring in a new field of science that would eventually be

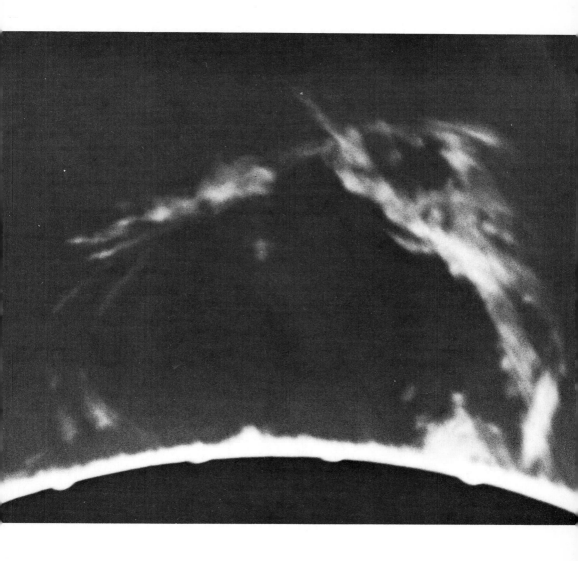

Figure 4-9.   A Prominence
Great flames of gas, called prominences, often extend tens of thousands of miles up into the sun's corona. This particular prominence rises to a height of 200,000 miles above the solar surface. Most prominences occur near sunspots. (Hale Observatories.)

79

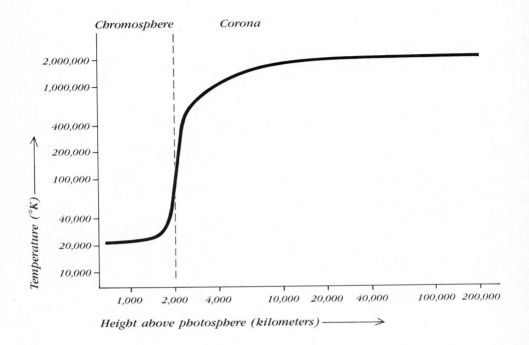

Figure 4-10. Temperature in the Solar Atmosphere
The higher you go in the sun's atmosphere, the hotter it is. The temperature of the photosphere (that is, the sun's "surface") is about 6,000 °K. In the chromosphere, the temperature has risen to about 20,000 °K. And in the corona, the temperature climbs to an incredible 2,000,000 °K.

called *nuclear physics.* Marie Curie and Ernest Rutherford showed that uranium atoms break down or "radioactively decay" into a host of other elements by emitting particles. In other words, contrary to nineteenth-century chemistry, it is possible to change one element into another.

In 1905, while attempting to resolve certain inconsistencies in electromagnetic theory, Albert Einstein formulated the *special theory of relativity.* Again contrary to nineteenth-century beliefs, this theory showed that it should be possible to convert matter into energy according to the now-famous equation $E = mc^2$.

And finally, just before the beginning of World War I, Rutherford and Bohr showed that atoms consist of small dense nuclei orbited by tiny electrons, like miniature quantum mechanical solar systems. These three new fields of physics — nuclear physics, relativity theory, and quantum mechanics — provided the tools for a three-pronged attack on the question of the sun's energy.

Hydrogen and helium together account for 98 percent of the sun's total mass. Perhaps for this reason, scientists turned to these two elements in trying to understand the source of the sun's energy.

In 1920, Sir Arthur Eddington proposed that the *fusion* of the nuclei of hydrogen atoms into helium was responsible for the sun's energy. The inspiration for this proposal is (in retrospect) perhaps obvious. If you look in any chemistry book, you find that one helium atom weighs nearly the same as four hydrogen atoms. Actually, four hydrogen nuclei weigh a tiny bit more than one helium nucleus. Accordingly, Eddington proposed the reaction

$$4H \longrightarrow He + energy$$

as the source of the sun's energy.

Eddington's reaction produces energy because the ingredients (four hydrogen nuclei) weigh more than the end product (one helium nucleus). The missing matter has been converted into energy according to $E = mc^2$. In this equation, m stands for the amount of missing matter, c is the speed of light, and E is the energy released. Since the speed of light is such a huge number (186,000 miles per second), a tiny amount of matter can result in enormous production of energy. Indeed, this conversion of matter into energy is far more powerful than any other processes ever discovered by mankind.

Over the next decade, Eddington's idea was elaborated upon. For example, we now know that there are intermediate steps in which the hydrogen nuclei combine in pairs. But the basic proposal was sound. The sun shines because 600 million tons of hydrogen are converted into helium *each second* at the sun's center. This process is often called hydrogen *burning* although nothing is burned in the usual sense.

Hydrogen burning occurs only at the sun's center. Only at the sun's center are the temperatures and pressures high enough to fuse the nuclei of hydrogen atoms together. At lower temperatures and pressures, the electric forces produced by the positive electric charges on the nuclei are strong enough to keep the nuclei apart. But at the sun's center, the nuclei collide with sufficient speed and violence that they can fuse in spite of the electric forces.

The conversion of 600 million tons of hydrogen into helium each second at the sun's center may sound like a prodigious rate of fuel consumption. But the sun contains so much hydrogen that this thermonuclear process can proceed comfortably for a total of 10 billion years. Geologists tell us that the age of the earth and the solar system is about 4.5 billion years. We therefore conclude that our star will continue to shine, without any major changes, for at least the next 5 billion years.

In order to understand why the sun shines, it was necessary to develop nuclear physics, quantum mechanics, and relativity theory. But the same understanding that explains that hydrogen is fused into helium at the sun's center also gave us the knowledge to build machines that fuse hydrogen into helium here on earth. Machines that fuse hydrogen into helium are called *hydrogen bombs*. The answer to the childlike question "Why does the sun shine?" directly provided mankind with the awesome knowledge of how to construct thermonuclear weaponry. The physical processes that produce the sunlight that makes the earth habitable can now be used to reduce our planet to a barren desolate rock. Whether we use our awesome knowledge and power for the betterment of humanity or the annihilation of our species is entirely a matter of our own free choice.

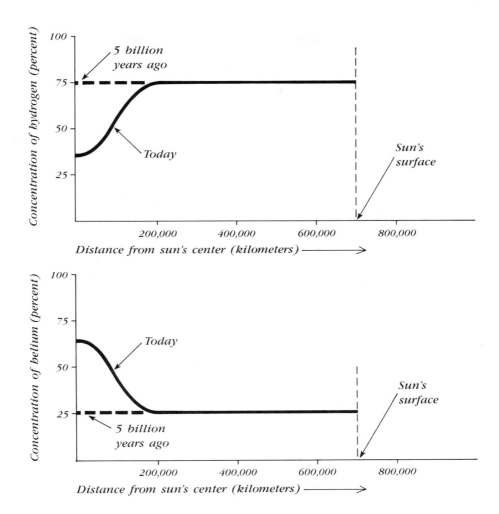

Figure 4-11.   *The Sun's Chemical Composition*
*Initially, the sun was 75 percent hydrogen and 25 percent helium
(plus a smattering of heavy elements). However, nearly 5 billion
years of thermonuclear reactions at the sun's center have depleted
the concentration of hydrogen and increased the amount of helium.*

5

The Birth of Stars

Galaxies are the largest objects in the universe. Although they have a wide range of shapes and sizes, typical galaxies (like our own) contain over a hundred billion stars and measure a hundred thousand light years in diameter. We are situated roughly two-thirds of the way between the center and edge of our own galaxy. The familiar Milky Way, which can be seen stretching across the sky on any dark moonless night, is our inside view of our galaxy.

Many galaxies have a characteristic pinwheel or spiral shape. Typical views of spiral galaxies are shown in Figures 4-1 and 5-1. We live in a spiral galaxy and the sun is between two of the spiral arms. Just as all the planets in the solar system go around the sun, all the stars in the sky are in orbit about the center of the galaxy. Our galaxy is rotating. It takes 250 million years for the sun to go once around the galaxy. Since the solar system is about 5 billion years old, we have circled the galaxy 20 times.

Contrary to first impressions, the spiral arms in spiral galaxies are *not* made out of stars. Those bright spots in Figure 5-1 that look like stars are not stars. They are huge clouds of glowing gas. The stars in a spiral galaxy are fairly evenly distributed throughout it.

Spiral arms in a galaxy are outlined by enormous clouds of glowing gas called *emission nebulas* or *H II regions*. These nebulas are millions of times bigger and brighter than any single star. Many astronomers believe that waves rippling through a galaxy compress the interstellar gas and dust along huge spirals arching outward from the galaxy's center. Emission nebulas and H II regions form in these locations where the waves have piled up the gas.

One beautiful emission nebula that can be seen with the naked eye is the middle "star" in Orion's sword. This nebula is shown in Figure 5-2. Another famous emission nebula is the Trifid Nebula, shown in Plate 2. It is interesting to note that there are pronounced dark regions in both of these nebulas.

Pronounced dark areas can often be found in close proximity to emission nebulas and H II regions. The famous Horsehead Nebula, shown in Plate 1, is a good example. For many years, astronomers believed that these dark areas were holes or voids in space where — for some reason — there were no stars, gas, or anything. In fact, almost 200 years ago, Sir William Herschel was viewing portions of the

Figure 5-1. A Spiral Galaxy
Many galaxies have a pinwheel or spiral shape. Almost every
starlike object that you can see in this photograph is really a huge
glowing cloud of gas. These clouds, called emission nebulas and
H II regions, outline the spiral arms. (Hale Observatories.)

Figure 5-2.   The Orion Nebula (also called M42 or NGC 1976)
*This beautiful emission nebula is just barely visible to the
naked eye as the middle "star" in Orion's Sword. This nebula
is just a small hot spot in a much larger cool cloud of interstellar gas. (Lick Observatory Photograph.)*

***Figure 5-3. The Doradus Nebula (also called
30 Doradus or NGC 2070)***
*This gigantic emission nebula is located in the Large Magellanic
Cloud. Sometimes called "the Tarantula," this nebula has a
luminosity equal to half a million suns and is thought to be 30
times larger than the Orion Nebula. (The Cerro Tololo
Inter-American Observatory.)*

89

Milky Way through his telescope when he suddenly cried out "My God. There is a hole in the sky!" Herschel had discovered the first *dark nebula.*

Dark nebulas are hard to find. When they are viewed through a telescope, they appear as regions of unusually few stars. A small dark nebula is shown in Figure 5-4. It is easy to overlook them completely.

Almost a hundred years after Herschel found the first dark nebula, astronomers began photographing vast regions of the Milky Way. These photographs soon revealed that it was far more likely that the dark regions were caused by obscuring material. Today we realize that cool clouds of gas and dust frequently block background stars and nebulosity.

Most emission nebulas and dark nebulas are found in the spiral arms of spiral galaxies. Of course, the bright emission nebulas are most easily noticed. But careful examination of photographs of spiral arms usually reveals lots of dark nebulosity beside the bright nebulas. The so-called *density waves* that compress gas and dust in spiral arms are responsible for creating the dark nebulas. As these waves ripple across a galaxy, they simply compress the cold interstellar material. Dark nebulas result at locations where an appreciable amount of cold interstellar gas and dust has been concentrated.

Think for a moment about one of these dark nebulas. It is reasonable to suppose that the nebula is not perfectly smooth, but rather contains some lumps. In other words, there are places in the nebula where a little extra gas and dust has piled up.

Think for a moment about one of these lumps inside a dark nebula. A lump is a lump because it contains a little more gas and dust than the surrounding regions. But because the lump has a little extra matter, it must also possess a slightly higher than average gravitational field. Due to the lump's greater gravity, it will attract some nearby gas and dust in the nebula. As the lump gathers gas and dust, its mass and gravity increase. This in turn allows the lump to attract still more gas and dust from the surrounding areas. The bigger it grows the more it attracts; the more it attracts the bigger it grows. By this self-perpetrating process, called *gravitational accretion*, a small lump can grow into a substantial object containing many solar masses

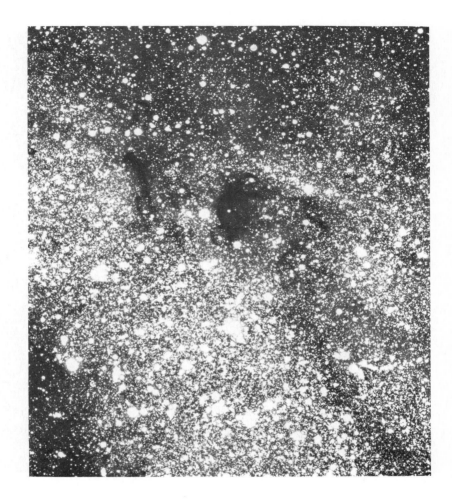

Figure 5-4.   A Dark Nebula
Thousands of stars are visible in this photograph of the Milky Way in Sagittarius. Notice the small, dark area near the center of the field of view. A cool, dark cloud of gas and dust is obscuring the background stars. (Lick Observatory.)

Figure 5-5.   Details of A Spiral Arm
Careful examination of spiral arms in spiral galaxies often reveals dark nebulosity right beside the bright emission nebulas. Enormous dark lanes of cool gas and dust stretching tens of thousands of light years are clearly visible along the inner edge of the spiral arm of this galaxy. (Hale Observatories.)

of interstellar material. These objects, called *globules,* are thought to be the embryos of stars.

Globules are very hard to find. They are much smaller, darker, and denser than the dark nebulas from which they form. As shown in Figure 5-6, globules can only be seen if they happen to be situated in front of a bright nebula. The smallest globules are only a trillion miles in diameter, roughly a hundred times bigger than our entire solar system.

Globules can be the birthplaces of stars. Because trillions upon trillions of tons of gas are pressing inward from all sides, the globule is unstable. Unable to support the weight of layer upon layer of gas, the globule begins to contract.

There is nothing to hold up all the gas, so a globule contracts fairly rapidly. In fact, at first the gas plunges toward the center of the globule from all sides. But soon pressures and densities at the globule's center become so great that contraction at the center is slowed. Contraction of the core stops when the internal gas pressures are strong enough to support the weight of the overlaying material. This is called *hydrostatic equilibrium.* Recent computations by Richard B. Larson at Caltech show that the center of a stellar embryo can achieve hydrostatic equilibrium even while the outer layers are still falling inward fairly rapidly. This creation of a fairly stable core signals the birth of a *protostar.*

The temperature inside a globule may start off at 50°K, or even cooler. But as the globule contracts to form a protostar, the temperature at the protostar's core may rise as high as 150,000°K. Quite naturally, as the gas is compressed under the influence of gravity, the temperature increases. Technically speaking, gravitational energy is converted into thermal energy. The final result is that the protostar has a hot center while the outer layers are still fairly cool.

Since the core of a protostar is much hotter than its surface, heat will start to flow outward from the center of the protostar. After all, heat always flows from warm places to cool places. Initially, heat is transported outward by *convection currents.* Blobs of hot material leisurely rise upward while equal amounts of cooler material sink downward, like a pot of soup simmering on the stove.

Hydrostatic equilibrium refers to a purely mechanical balance: at every point, the pressure outward equals the pressure inward.

Figure 5-6. Globules
Numerous globules are seen silhouetted against the bright background nebulosity. Globules represent the earliest stages in the birth of stars. In only a few million years, this region of space will be covered with dazzlingly bright newborn stars. (Lick Observatories.)

Quite simply, hydrostatic equilibrium means that a star or protostar can hold up its own weight. But in order for a star or protostar to be stable, there must also be a thermal balance: at every point, the temperature must be just high enough to keep energy flowing outward. There are primarily two ways in which energy is transported outward in a star: convection and radiation. Convection currents in the protostar provide the necessary thermal balance. The protostar is therefore said to be in *convective equilibrium.*

As convection transports energy outward away from the protostar's core, the pressure and temperature at the core start to drop. But as soon as the pressure begins to fall, the enormous weight of all the gas above the core presses inward more than ever. This causes the core to become even more compressed. But this added compression produces still further heating of the gases. The final result is that temperature and pressure at the protostar's center become greater than before. In short, the outward flow of energy results in a gradual contraction of the protostar. And the gradual contraction of the protostar keeps the heat flowing outward.

Eventually, the pressures and densities inside the protostar become so great that convection is no longer an efficient means of transporting energy. Blobs of hot and cool material cannot easily rise and fall anymore. Then the convection currents at the core cease and energy is transported outward primarily through radiation. The transition from convection to radiation as the primary means of energy transport signals a major change in the structure of the protostar. Initially *radiative equilibrium* is established at the protostar's center. But this region of radiative equilibrium slowly grows as the gradual contraction of the protostar continues to provide the protostar's core with thermal energy.

Finally, the pressures, temperatures, and densities at the center of a slowly contracting protostar have become so great that the nuclei of hydrogen atoms can be fused. When the temperature in the core has risen to about 4 million degrees Kelvin, hydrogen nuclei are packed so tightly and are moving so rapidly that they can collide and stick together. The ignition of this thermonuclear reaction, called hydrogen burning, is the final event in the birth of a star. The protostar no longer has to rely on gravitational contraction for its source of thermal energy. Thermonuclear fusion now provides the energy. Contraction stops. A star is born.

Astronomers see many clusters of newborn stars in the sky. A particularly beautiful example is shown in Figure 5-7. The wide-angle view in Figure 5-8 reveals extensive dark nebulosity and emission nebulosity surrounding a cluster of young stars called NGC 2264. It is believed that nuclear reactions are just beginning in the centers of these stars.

This description of the birth of a star is the result of many years of work by many astrophysicists, each contributing his own theoretical or mathematical insight to the problem. We imagined a huge ball of gas much bigger than the entire solar system that contains several solar masses of cool gas and dust. Using all the laws of physics, a great deal of mathematics, and a computer to speed the calculations, we simply asked what happens as the ball of gas contracts under the influence of gravity. Our main conclusions were presented on the preceding few pages.

Astrophysicists think they are on the right track with their ideas about the birth of stars because the computer's answers agree with observations. At any point in our calculations we can ask the computer to tell us the luminosity and surface temperature of the contracting protostar. Once we know the luminosity and surface temperature, we can draw a dot on the Hertzsprung-Russell diagram. Since the protostar is shrinking in size while its core is heating up, the luminosity and surface temperature are constantly changing. The dot representing a protostar on the H-R diagram is wandering. The path that the dot follows is called an *evolutionary track.* The obvious question is: Where do all the dots — each representing a protostar — end up when hydrogen burning is finally ignited?

The evolutionary tracks of five representative protostars are shown in Figure 5-9. These tracks were initially deduced by the Japanese astrophysicist Chushiro Hayashi in 1965. By 1969, Richard B. Larson elaborated significantly on Hayashi's ideas and worked out all the details.

During the initial stages in the contraction of a protostar, heat is transported from the center to the surface by convection. As long as the protostar is in convective equilibrium, its dot on the H-R

Figure 5-7. A Stellar Maternity Ward
*Globules and newborn stars are side by side in this nebulosity in
the constellation of Serpens. The hot, bright stars that form first are
responsible for lighting up the nebula. (Lick Observatory
Photograph.)*

Figure 5-8. A Nursery of Stars
Both bright and dark nebulas are visible in this wide-angle photograph of a region in Monoceros. Bright emission nebulosity dominates the upper half of the photograph. Dark nebulosity is mostly around the center and lower third of the photograph. (Copyright by the National Geographic Society —Palomar Observatory Sky Survey. Reproduced by permission from the Hale Observatories.)

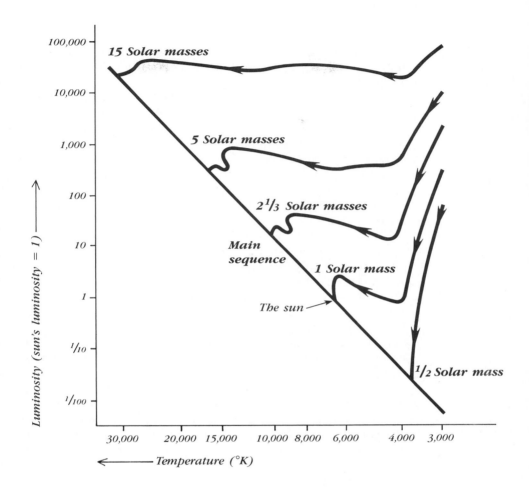

Figure 5-9. Pre-Main Sequence Evolution
The evolutionary tracks of five protostars are shown on this H-R diagram. High-mass protostars evolve very rapidly and take only a few thousand years to arrive at the main sequence. Low-mass protostars contract much more slowly; hundreds of millions of years pass before hydrogen burning is ignited at their cores.

diagram moves straight down. The luminosity falls as the protostar shrinks, but the surface temperature remains fairly constant at about 4,000°K.

The second stage in the contraction of a protostar occurs when convection ceases at its core and energy is transported outward by radiation. When this happens, the dot on the H-R diagram turns sharply to the left and starts moving toward the main sequence. This means that the luminosity has stopped declining and now the surface layers of the protostar are heating up. Notice, incidentally, for a very low mass star (for example, ½ solar mass) that the evolutionary track never turns to the left. Such stars are always fully convective.

The third and final stage in the birth of a star occurs when the core of the protostar becomes hot enough to ignite thermonuclear reactions. When this happens, the dots representing the protostars have almost arrived at the location of the main sequence. Notice that the evolutionary tracks dip a little bit as the structure of the star adjusts to the onset of hydrogen burning. The result is that all the dots end up on the main sequence.

This reveals the true meaning of the main sequence. A newborn star is a main sequence star. All main sequence stars are powered by hydrogen burning at their cores.

Also notice that the dots end up in the right places on the main sequence. After all, from the mass-luminosity relation, low-mass main sequence stars are dim, whereas high-mass main sequence stars are bright. Tracks for low-mass protostars go to the cool, dim end of the main sequence while tracks for high-mass protostars go to the hot, bright end of the main sequence. The track for a one solar mass protostar terminates in the middle of the main sequence at a luminosity of one sun and a surface temperature of about 6,000°K.

In following theoretical stars along these travels, one important fact emerges. The more massive the star, the more rapidly it evolves. High-mass stars very rapidly build up the necessary temperatures and pressures to ignite hydrogen burning at their centers and therefore take a short time to get to the main sequence. Low-mass stars take a much longer time. The following table tells the time it takes for protostars to reach the main sequence.

Mass of protostar (sun = 1)	Time to reach main sequence
30 solar masses	30 thousand years
10 solar masses	300 thousand years
4 solar masses	1 million years
2 solar masses	8 million years
1 solar mass	30 million years
$\frac{1}{2}$ solar mass	100 million years
$\frac{1}{5}$ solar mass	1 billion years

Theoretical calculations about the births of stars go hand in hand with observations of young clusters of stars. For example, the cluster NGC 2264, shown in Figure 5-10, looks as if it contains many young stars because of all the nebulosity around it. Indeed, when we measure the brightnesses and colors of these stars and plot the resulting data on an H-R diagram, most of the dots fall very near the main sequence. But since the dots have not arrived at the main sequence (see Figure 5-12), we conclude that this cluster contains many stars that are only now igniting hydrogen burning at their cores.

A slightly more mature cluster of young stars is the Pleiades shown in Figure 5-11 and Plate 4. When the observations of brightnesses and colors are plotted on an H-R diagram, the dots fall right along the main sequence. All of these stars are comfortably burning hydrogen at their centers.

Eventually all the hydrogen at the center of a main sequence star is used up. This causes major and dramatic changes in the star's structure. These later stages of stellar evolution produce a wide range of objects and phenomena that astronomers are just beginning to discover and to understand.

Figure 5-10.   The Cluster/Nebula NGC 2264
This cluster in Monoceros contains many very young stars. Some are in the final stages of gravitational contraction, while others have just started to burn hydrogen at their cores. See Figure 5-8 for a wide-angle view of this part of the sky. (Hale Observatories.)

Plate 1.   Horsehead Nebula (also called NGC 2024) in Orion.
Copyright by the California Institute of Technology and the Carnegie Institution of Washington. Reproduced by permission from the Hale Observatories.

Plate 2.   Trifid Nebula (also called M 20 or NGC 6514) in Sagittarius.
Copyright by the California Institute of Technology and the Carnegie Institution of Washington. Reproduced by permission from the Hale Observatories.

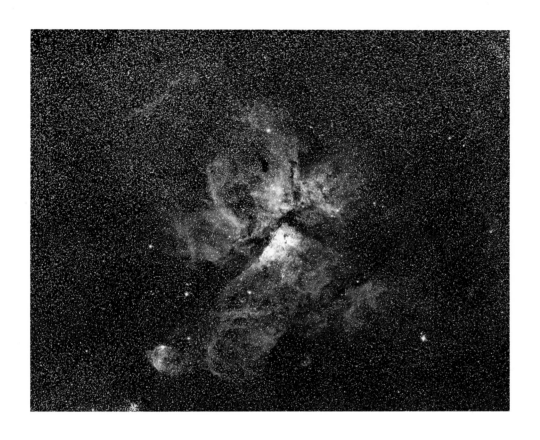

***Plate 3. Eta Carinae Nebula (also called
NGC 3372) in Carina.***
*Copyright by the Association of Universities for Research in
Astronomy, Inc. The Cerro Tololo Inter-American Observatory.*

Plate 4. Pleiades (also called M 45 or NGC 1432) in Taurus.
Copyright by the California Institute of Technology and the Carnegie Institution of Washington. Reproduced by permission from the Hale Observatories.

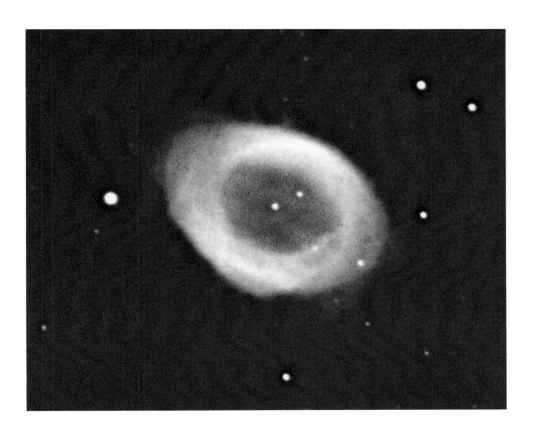

***Plate 5.   Ring Nebula (also called M 57 or
NGC 6720) in Lyra.***
*Copyright by the California Institute of Technology and the
Carnegie Institution of Washington. Reproduced by permission
from the Hale Observatories.*

***Plate 6. Dumbbell Nebula (also called M 27 or
NGC 6853) in Vulpecula.***
*Copyright by the California Institute of Technology and the
Carnegie Institution of Washington. Reproduced by permission
from the Hale Observatories.*

Plate 7.   Veil Nebula (also called NGC 6992) in Cygnus.
Copyright by the California Institute of Technology and the Carnegie Institution of Washington. Reproduced by permission from the Hale Observatories.

***Plate 8.   Crab Nebula (also called M 1 or
NGC 1952) in Taurus.***
*Copyright by the California Institute of Technology and the
Carnegie Institution of Washington. Reproduced by permission
from the Hale Observatories.*

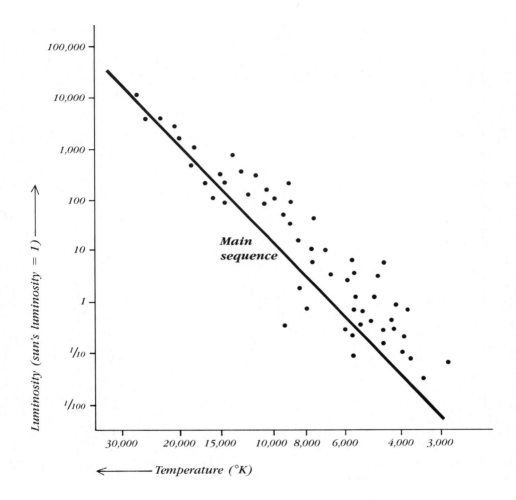

**Main
sequence**

Figure 5-11.   NGC 2264 on the H-R Diagram
*By measuring the brightnesses and colors of the stars in NGC 2264,
astronomers can plot the cluster on an H-R diagram. Each dot
represents a star. The positions of the dots reveal that the cluster is
full of protostars in the final stages of contraction or new stars that
have just ignited hydrogen burning at their cores.*

Figure 5-12.   The Pleiades (also called M45 or NGC 1432)
This cluster of young stars can be seen clearly with the naked eye on winter nights. Sometimes called "The Seven Sisters," it is in the constellation of Taurus, the bull. See Plate 4 for a color photograph of this cluster. (Lick Observatory Photograph.)

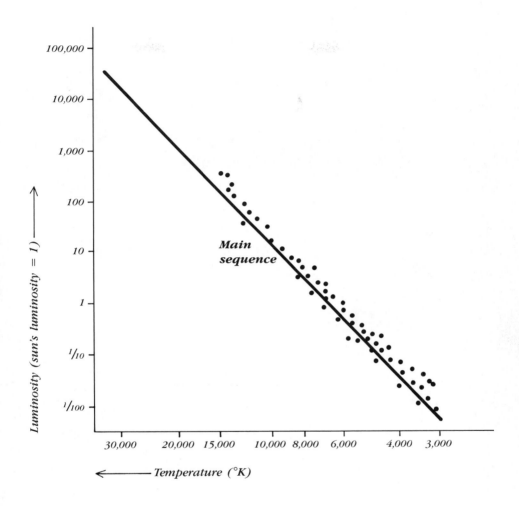

Figure 5-13.   The Pleiades on the H-R Diagram
By measuring the brightnesses and colors of the stars in the Pleiades, astronomers can plot the cluster on an H-R diagram. Each dot represents a star. Since all the dots lie along the main sequence, all the stars in the cluster are burning hydrogen at their centers.

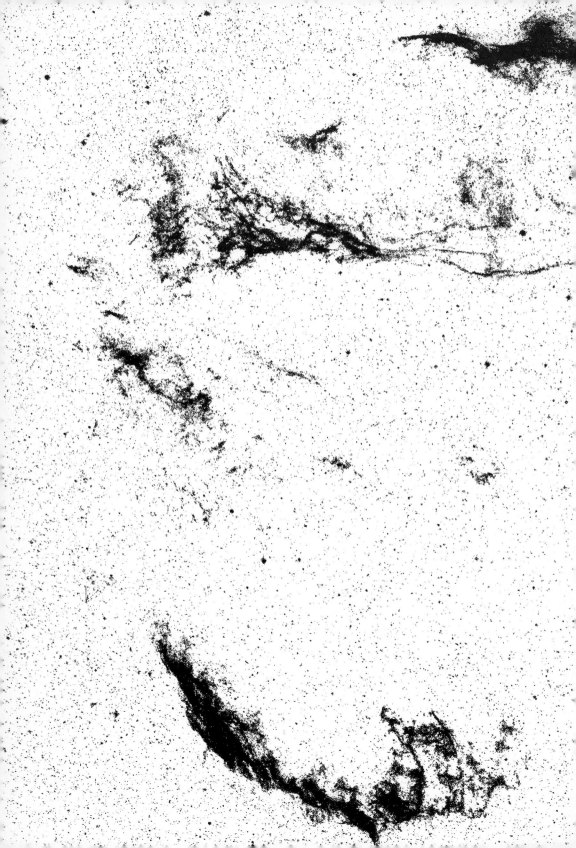

6
Maturity and Old Age

Gazing up into the nighttime sky you can see hundreds upon hundreds of stars scattered across the heavens. It is truly remarkable that most of these stars have a very great deal in common. The vast majority are main sequence stars. In other words, if you measured the star's surface temperature and its luminosity, the dot representing that star on a Hertzsprung-Russell diagram would be somewhere along the main sequence. Equally important, all main sequence stars are newborn stars. These stars have only recently formed from interstellar clouds of gas and dust. Hydrogen burning has only recently been ignited at the centers of these stars. These facts illustrate the true meaning of the main sequence. Main sequence stars are young stars that are burning hydrogen at their cores.

Hydrogen burning obviously cannot go on forever at the center of a star. For example, the sun converts 600 million tons of hydrogen into helium each second. After a total of 10 billion years, the hydrogen at the sun's core will be completely used up. At that time, 5 billion years into the future, the sun will evolve dramatically into a radically new kind of star.

All main sequence stars eventually deplete the hydrogen fuel at their centers. The rate at which they consume this fuel depends critically on their masses. Low-mass stars burn hydrogen very slowly, whereas high-mass stars devour it at a furious rate. Quite simply, massive stars easily develop very high pressures and temperatures at their centers because of the enormous weight of all the overlying material. Thermonuclear reactions readily occur under these conditions of high temperature and pressure. Thus, although high-mass stars contain lots of hydrogen fuel, they burn this fuel at a much faster rate than the low-mass stars.

The following table lists the approximate times that stars of different masses spend on the main sequence. These lifetimes on the main sequence tell us how long it takes for stars to consume all the hydrogen in their cores.

A star on the main sequence is comfortably burning hydrogen at its center. Nothing dramatic happens as this thermonuclear reaction gradually alters the chemical composition in the core of the star. As the hydrogen is depleted, the amount of helium increases. The structure of the star easily adjusts to these slow changes in chemical

Surface temperature (°K)	Luminosity (sun = 1)	Mass (sun = 1)	Time on main sequence
35,000	80,000	25	3 million years
30,000	10,000	15	15 million years
11,000	60	3	500 million years
7,000	5	1½	3 billion years
6,000	1	1	10 billion years
5,000	½	¾	15 billion years
4,000	¹⁄₄₀	½	200 billion years

composition. Over millions of years, the core contracts slightly, thereby turning some gravitational energy into thermal energy. In order to maintain hydrostatic and thermal equilibrium, the outer layers of the star expand slightly. This means that the star gets a little brighter and its atmosphere gets a little cooler. In short, the late stages of hydrogen burning at the core of a main sequence star are marked by a small increase in the star's luminosity and a slight decline in the star's surface temperature. Over the next few billion years, the sun will get a little brighter and perhaps a few degrees cooler.

Hydrogen burning on the main sequence is rather uneventful. During this time, a star passes from adolescence into adulthood. But when all the hydrogen reserves at its center are exhausted, the star starts to change rapidly and dramatically.

As soon as all the hydrogen at the center of a main sequence star is used up, hydrogen burning stops. For millions or even billions of years, the star had relied on hydrogen burning to maintain hydrostatic and thermal equilibrium. But when the thermonuclear reactions shut off, equilibrium in the star's core is severely disrupted. The core can no longer support itself. Under the overwhelming influence of gravity, the star's core rapidly contracts. This core contraction converts lots of gravitational energy into thermal energy that pushes the

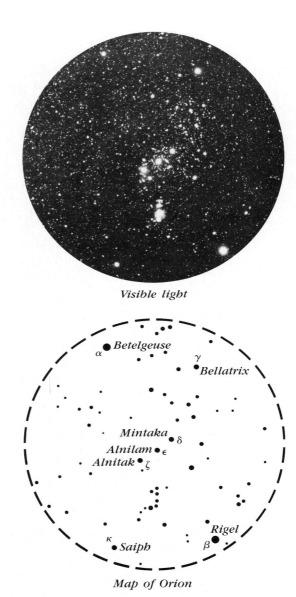

Visible light

Map of Orion

Ultraviolet light

Figure 6-1. Massive, Young Stars in Orion
The constellation of Orion is full of hot, young stars. The hottest stars (surface temperatures between 10,000 and 50,000 °K) can be picked out most easily in the ultraviolet view because they emit primarily ultraviolet light. All of these young, massive, brilliant stars are rapidly consuming hydrogen at their cores. They will survive as main sequence stars for only a few million years. (Courtesy of G. R. Carruthers, N.R.L.; Hale Observatories.)

star's atmosphere outward. In short, the cessation of core hydrogen burning causes the star's core to contract and its atmosphere to expand.

As the burned-out, helium-rich core collapses, the pressure and temperature at the center of the star rise rapidly. This heats the layers above the core. But these layers in the star—between the core and the surface—still contain plenty of fresh hydrogen. After a relatively short period of time, temperatures above the contracting core reach 4 million degrees Kelvin. This is so hot that hydrogen burning ignites in a shell surrounding the core. The ignition of *shell hydrogen burning* surrounding an inactive helium-rich core promptly produces major changes throughout the rest of the star.

With the ignition of shell hydrogen burning, a star suddenly has a new thermonuclear source of energy. The continued contraction of the star's inert core and this new outpouring of energy cause the star to expand enormously. The star's outer layers are pushed farther and farther outward as the structure of the star tries to maintain equilibrium with its new energy source.

An important concept in thermodynamics is that anytime a gas expands, it cools. This is simply how normal gases behave. Compress some gas and it gets hot (diesel engines use this principle). Allow some gas to expand and it gets cool (refrigerators and air conditioners use this principle).

As the energy from the contracting core and shell hydrogen burning cause a star to expand, the atmosphere of the star starts to cool. Regardless of where on the main sequence the star originally came from, the star's surface temperature eventually gets as low as 4,000°K. According to Wien's law, an object at 4,000°K emits reddish light primarily. The star has become very large and very red. The star has become a red giant!

Perhaps one of the best ways to display the metamorphosis of main sequence stars into red giants is to follow their *evolutionary tracks* on a Hertzsprung-Russell diagram. One of the pioneers in calculating post-main sequence evolution is Icko Iben. The results of his calculations are shown in Figure 6-2, in which the evolutionary tracks of six stars are displayed.

As I mentioned before, the late stages of core hydrogen burning are characterized by a slight and gradual expansion of the star.

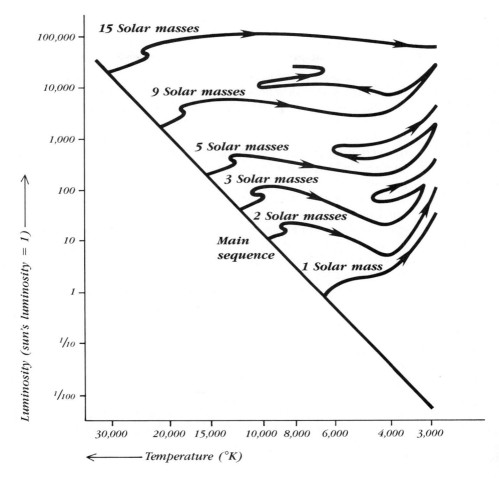

Figure 6-2. Post-Main Sequence Evolution
The evolutionary tracks of six middle-aged stars are shown on this
Hertzsprung-Russell diagram. The cessation of core hydrogen
burning is marked by a small kink in the paths followed by
moderate- and high-mass stars as they start to evolve away from the
main sequence. The stars then race across the H-R diagram as shell
hydrogen burning is ignited above their cores. They become
full-fledged red giants when helium burning is ignited at their
centers.

During this phase, the evolutionary tracks lead in the direction of increased luminosity (because the stars get bigger) and decreased surface temperature (because the stars' atmospheres expand slightly and therefore cool). The dots representing hypothetical stars on a H-R diagram slowly move away from the main sequence.

As soon as hydrogen burning shuts off at the center of an old main sequence star, a small but pronounced kink usually appears in the evolutionary track, as shown in Figure 6-2. This signals major changes inside the star in response to the termination of core hydrogen burning.

As the inert core contracts and shell hydrogen burning begins, the dot representing a star moves very rapidly across the H-R diagram. As shown in Figure 6-2, massive stars maintain a fairly constant luminosity during this transition phase. The diameters of these stars are dramatically increasing while their surface temperatures are plunging. The combined effect of an increase in size (which alone would cause a star to get bright) along with a substantial decline in temperature (which alone would cause a star to get dim) results in a fairly constant luminosity.

While the outer layers of the star are expanding and cooling, the inert core continues to become very compressed. This means that the densities and temperatures deep inside the star have simply been getting higher and higher. Finally, when the core temperature reaches 100 million degrees Kelvin, a new thermonuclear reaction is ignited. The temperature and pressure are now so high that helium nuclei can be fused together to form carbon and oxygen according to the reaction,

$$3He \rightarrow C$$
$$\text{and} \quad C + He \rightarrow O$$

This is *helium burning.*

Immediately before the ignition of core helium burning, the evolutionary tracks of moderate and low-mass stars have turned upwards on the H-R diagram. These stars have been getting bigger and brighter, but have maintained a fairly constant surface temperature around 4,000°K. But as soon as helium burning is ignited, the evolutionary track turns down and to the left on the H-R diagram. This leads the star part of the way back towards the main sequence.

For stars of moderate and high mass (that is, greater than 2 solar masses) core helium burning ignites in a comfortable fashion as the necessary temperatures are achieved at the stars' centers. As these stars adjust to core helium burning surrounded by shell hydrogen burning, their evolutionary tracks begin to wander on the red giant region of the H-R diagram.

For low-mass stars (that is, less than about 2 solar masses) the ignition of helium burning is sudden and explosive. When helium burning finally turns on, a thermal runaway called the *helium flash* occurs. This explosive thermonuclear event is confined completely to the star's core: nothing dramatic happens at the star's surface. Astronomers will never actually see a star experiencing the helium flash. They "see" the flash only in their computer calculations of the star's internal structure. After the flash, the star settles down to core helium burning surrounded by shell hydrogen burning.

Almost every reddish star you can see in the night sky is a red giant burning both hydrogen and helium far below its surface. Aldebaran (in Taurus), Antares (in Scorpius), Arcturus (in Boötes), and Betelgeuse (in Orion) are familiar red giants. But in addition to these well-known examples, astronomers find many red giants in old clusters of stars.

Globular clusters are the oldest clusters of stars that we have seen in the sky. They are believed to have formed billions upon billions of years ago from cool primordial clouds of gas. Figure 6-3 shows a good example of one of these ancient star clusters. When the astronomer carefully measures the brightnesses and temperatures of the stars in this cluster and plots the data on an H-R diagram, few dots lie along the main sequence. Instead, many dots fall in the red giant region, as shown in Figure 6-4.

From examining the data for globular clusters such as that in Figure 6-4, we realize that only dim, low-mass stars still remain on the main sequence. As expected, the more massive stars have all evolved away from the main sequence. Once again we see that the more massive a star is, the more rapidly it evolves.

Figure 6-3. **The Globular Cluster M3 (also called NGC 5272)**
Globular clusters, like this one in the constellation of Canes Venatici, typically contain over 100,000 stars. The most ancient stars in the sky are in globular clusters. Astronomers are sure that, because of the total absence of interstellar gas and dust, star formation in globular clusters stopped billions of years ago. (Lick Observatory Photograph.)

116

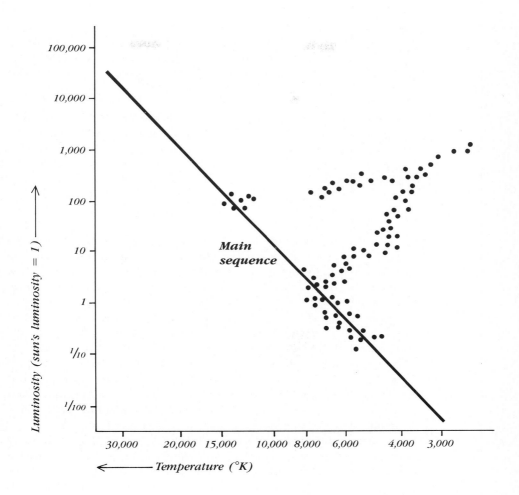

Figure 6-4.   The Globular Cluster M3 on the H-R Diagram
*By measuring the brightnesses and surface temperatures of the
stars in the globular cluster, the astronomer can plot the data on a
H-R diagram. Each dot represents a star. Notice that there are
many red giants, whereas only the low-luminosity, low-mass end of
the main sequence is intact. Also notice how the positions of the
dots correlate with the evolutionary tracks in Figure 6-2.*

Thermonuclear process	Minimum star mass required (sun = 1)	Approximate ignition temperature (°K)	Typical operating temperature (°K)
Hydrogen burning	$1/10$	4 million	20 million
Helium burning	$1/2$	100 million	200 million
Carbon burning	4	600 million	800 million
Oxygen burning	6	1 billion	$1\frac{1}{2}$ billion
Silicon burning	8	2 billion	$3\frac{1}{2}$ billion

Eventually, all the helium at the core of a red giant has been burned up and converted into carbon and oxygen. Obviously, the helium burning shuts off and the inert core starts to contract. What happens next critically depends on the mass of the star. As shown in the above table new thermonuclear reactions can be ignited *if* the temperature gets high enough. The star's mass is the deciding factor.

Stars that have burned all the helium in their cores are rapidly approaching old age. They are the senior citizens of our galaxy. As the inert, carbon-rich core contracts, temperatures above the core soon get high enough to ignite *shell helium burning*. A star at this late stage in its life has thermonuclear reactions occurring in *two* shells: a hydrogen burning shell surrounding a helium burning shell. The structure of one of these stars is shown in Figure 6-5.

If a star contains more than 4 solar masses, then densities and temperatures at its center get high enough to ignite *carbon burning*. The thermonuclear fusion of carbon nuclei produces oxygen, neon, sodium, and magnesium. Three independent thermonuclear reactions are therefore occurring in the star: core carbon burning surrounded by shell helium burning surrounded by shell hydrogen burning.

Eventually, of course, all of the carbon at the center of one of these stars is burned up. Carbon burning shuts off, the star's core contracts, and *shell* carbon burning is ignited. If the star is more massive than about 6 solar masses, temperatures at the core climb

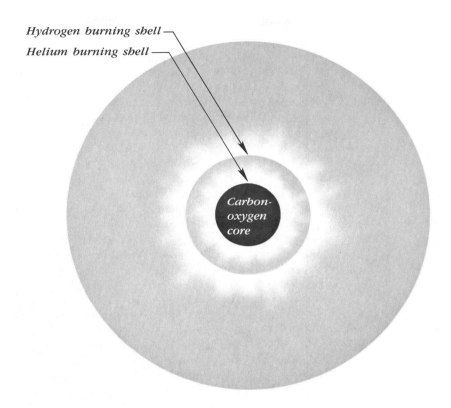

Hydrogen burning shell —

Helium burning shell —

Carbon-oxygen core

Figure 6-5.   The Structure of an Old Low-Mass Star
This cross-section diagram shows double shell burning in a very old, low-mass star. The inert core contains mostly carbon and oxygen. For stars less than 4 solar masses, temperatures never get high enough to ignite any additional nuclear reactions. At this stage, the star has finished its life cycle and is just about to die. (Diagram is not to scale.)

119

above 1 billion degrees. At these extraordinary temperatures, *oxygen burning* commences. This reaction produces silicon and sulfur.

The cessation of core oxygen burning is followed by yet another core contraction and the ignition of shell oxygen burning. And if the star is sufficiently massive enough to push the central temperature above 2 billion degrees, *silicon burning* is ignited.

Nickel and iron are the ashes of silicon burning. Unlike the lighter elements, iron does not "burn." The buildup of iron at the core of a very massive star (for example, greater than about 10 solar masses) signals the impending, violent death of the star. Immediately before its death, a massive star can burn silicon in a shell surrounding the iron-rich core, as shown in Figure 6-6.

A massive star that has reached this stage in its life cycle is in imminent danger of blowing itself apart. The burned-out iron-rich core becomes unstable and implodes. This sudden gravitational collapse releases an enormous amount of energy that completely tears the star apart, producing a *supernova*. During a supernova explosion, a bewildering array of exotic nuclear reactions occur. Any element that has not been produced during earlier thermonuclear processes is now created in a few moments as the star rips itself apart.

Astronomers have good reason to believe that hydrogen and helium were the only elements present in the primordial universe. Only hydrogen (75 percent) and helium (25 percent) could have arisen from the inferno of the primordial fireball of the creation event 20 billion years ago. Yet the world around us is filled with dozens of different elements. Carbon, oxygen, nitrogen, calcium, potassium, and phosphorous — to name a few — are important in all plants and animals here on earth.

When a dying star becomes a supernova, it violently ejects vast quantities of matter out into space. Although the star was originally composed almost entirely of hydrogen and helium, nuclear processes have converted a large fraction of these substances into all the heavier elements. A supernova explosion therefore spews large amounts of heavy elements out into the universe. Interstellar clouds of gas become enriched with these heavy elements. Stars and planets that condense from this enriched interstellar medium will contain all these elements.

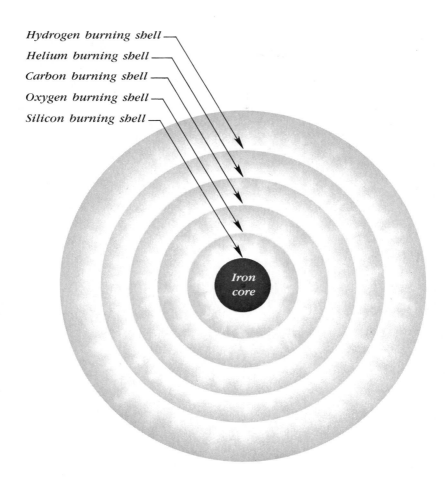

Hydrogen burning shell

Helium burning shell

Carbon burning shell

Oxygen burning shell

Silicon burning shell

Iron core

Figure 6-6. The Structure of an Old High-Mass Star
This cross-section diagram shows multiple shell burning in a very old, massive star. The creation of a burned-out, iron-rich core signals the impending, catastrophic death of the star. A star at this final stage in its life cycle is just about to blow itself apart and become a supernova. (Diagram is not to scale.)

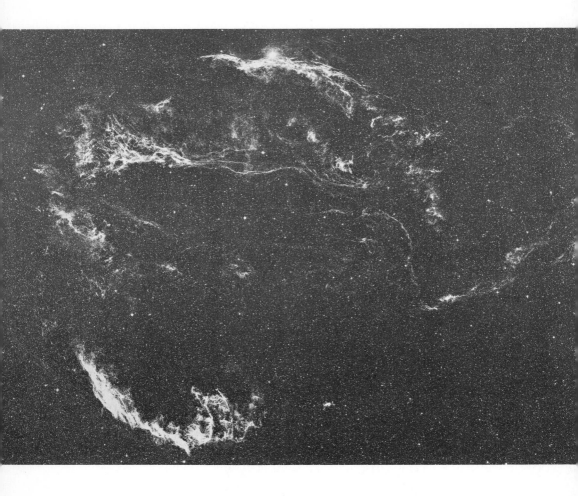

Figure 6-7. A Supernova Remnant
This nebula in the constellation of Cygnus is the result of a star that became a supernova about 50,000 years ago. During a supernova explosion, a star can violently eject a large fraction of its mass. In doing so, the dying star enriches the interstellar medium with heavy elements that were created deep within the star's interior. (Hale Observatory.)

The sun is a late-generation star. Fifteen billion years had passed before it even began to form. During that time, countless millions of massive stars lived out their lives and supplied our region of space with the full array of chemical elements. Our world is made out of these chemical elements. Our world and ourselves—every atom in our bodies, every atom we touch and eat and breathe—were created long ago at the centers of doomed stars. Quite literally, we are made of the dust of stars.

7

The Deaths
of Stars

Something happened here 5 billion years ago. Perhaps a spiral arm of our galaxy passed through our region of space. Or perhaps a nearby massive star became a supernova. In either case, the final result would have been the same: The sparce interstellar medium was subjected to a compression. This slight compression of the near-perfect vacuum of our region of space was sufficient to start the formation of the sun. Within a very short time — less than 100 million years — our star was born.

Our sun has already lived out half of its life as a main sequence star. Sufficient fuel now remains for only 5 billion years of core hydrogen burning. Five billion years from now, when hydrogen burning shuts off, the sun's core will contract while its atmosphere expands and cools. As the temperatures around the contracting core rise, shell hydrogen burning will be ignited. In a relatively short time, the sun will grow so large that it will swallow the earth. The oceans will boil, the continents will melt, and our planet will be vaporized. At the time of the helium flash, the red giant sun will be slightly larger than the earth's orbit.

After a few billion years, all the helium at the sun's center will have been converted into carbon and oxygen. The cessation of core helium burning is followed by a second core contraction and the ignition of shell helium burning. All this time, the hydrogen burning shell will have been gradually moving outward in search of fresh fuel.

Low-mass stars like our sun never can develop the necessary temperatures at their cores to ignite any new thermonuclear fires. Double shell burning is therefore the final stage in the life cycle of a star like the sun. The burning of hydrogen and helium in two concentric shells surrounding a carbon-oxygen core signals the imminent death of our star.

All stars are constantly shedding matter throughout their lives. The sun, for example, emits high-speed particles (protons and electrons) in the form of a *solar wind.* Indeed, it is reasonable to suppose that every star is surrounded by a *stellar wind* that consists of particles spewing outward from the star's upper atmosphere. During the course of its life, a star might shed half of its mass through the stellar wind, which can — at times — blow with great fury. Satellite observations reveal that the brightest red giants are losing large quantities of gas in this fashion. Such stars are so huge and distended that their outer atmospheres are held on only loosely.

As a low-mass star burns helium and hydrogen around an inert, carbon-oxygen core, thermal instabilities can develop. The star begins to pulsate, thereby causing its atmosphere to expand briefly. These pulses, which are typically separated by only a hundred thousand years, gradually get stronger and stronger. Eventually the pulses become so powerful that the outer layers of the star completely separate from the core. As the star's envelope gently moves outward into space, the very hot, very dense core is exposed.

As astronomers search the skies, they find evidence of many low-mass stars that have gently ejected their outer envelopes. The expanding envelopes are *planetary nebulas*. Excellent, well-known examples include the Ring Nebula (see Figure 7-1 and Plate 5) and the Dumbbell Nebula (see Plate 6). Notice that each nebula is clearly concentric about one particular star. This "star" is actually the burned out carbon-oxygen core that has finally been exposed to view. Its surface temperature is extremely high, typically 100,000°K. Our knowledge of Wien's law tells us that anything at this temperature must be emitting vast quantities of ultraviolet radiation. This outpouring of energetic ultraviolet light excites the atoms in the expanding shell of gas, which allows us to see planetary nebulas. Without the hot central star, the expanding gases would simply cool off and become invisible. But instead, radiation from the burned out core keeps the gases glowing. The resulting nebulas are among the most beautiful objects in the sky.

Planetary nebulas have a very temporary existence. After a very short period of time — perhaps 50,000 years — the expanding envelope of gas has become so thin and dispersed that it no longer glows. The gases simply merge with the interstellar medium. All the planetary nebulas that we see are therefore very recent phenomena; they must be less than 50,000 years old. Astronomers have found about a thousand of these nebulas across the heavens.

As the expanding gases of a planetary nebula begin to dissipate, the hot central star starts to cool off. There is no possibility that

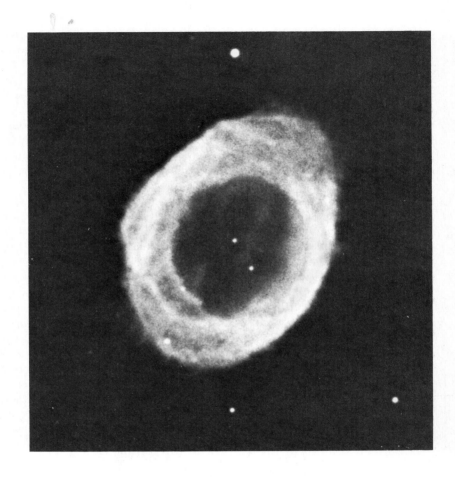

Figure 7-1.   The Ring Nebula (also called M57 or NGC 6720)
*This planetary nebula in Lyra is one of the most beautiful sights
in the sky. About 20,000 years ago, a star was nearing the end of
its life. Its outer atmosphere was ejected and the hot dense core was
exposed. The "star" exactly at the center of the ring is now
contracting to become a white dwarf. (Lick Observatory
Photograph.)*

128

carbon burning would be ignited inside one of these low-mass stars. The star's surface temperature therefore simply declines as energy is radiated into space.

As the outer layers of the burned-out star begin to cool, the star's core becomes more and more compressed. Trillions upon trillions of tons of matter pressing inward from all sides cause the densities inside the star to become higher and higher. Finally, when the density reaches about 1,000 tons per cubic inch, all the electrons inside the star can act together to provide a very strong pressure that powerfully resists any further compression. This unusual pressure arises from the quantum mechanical properties of electrons first described by Wolfgang Pauli in the 1920s. As a direct consequence of the *Pauli Exclusion Principle,* there is a minimum volume into which a number of electrons may be squeezed. Beyond that stage, the electrons become highly incompressible and they vigorously resist any further squeezing. This behavior of matter is called *degeneracy* and the pressure that resists further compression is known as *degenerate electron pressure.*

This is not the first time that matter has become degenerate inside the star. For example, immediately before the helium flash, the star's core was degenerate. But now, instead of being an interesting detail in the star's structure, degeneracy dominates it.

When the contracting central star in a planetary nebula has shrunk to the size of the earth (that is, roughly 10,000 miles in diameter), degenerate electron pressure throughout the star's interior has become so great that the star stops contracting. Degenerate electron pressure has become sufficient to support trillions upon trillions of tons of burned-out stellar material. By this time, the surface temperature of the star has fallen below 50,000°K.

The star is dead — there are no internal sources of energy. The star is small — typically, an entire solar mass of burned-out stellar matter has been compressed into a sphere no larger than the earth. The star is hot. Although it may start with a surface temperature of 50,000°K, the star cools very slowly by emitting radiation primarily at ultraviolet wavelengths. Consequently, as seen through a telescope, one of these tiny stars appears to be "white hot." It has become a *white dwarf.*

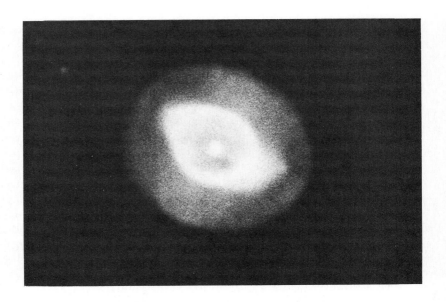

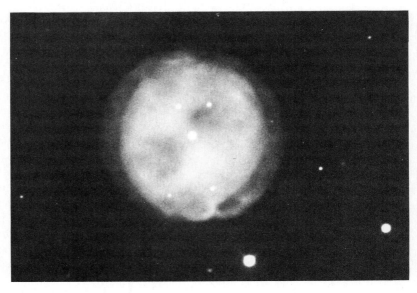

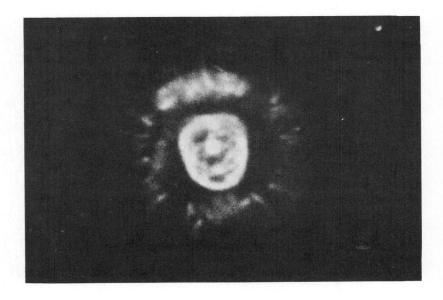

Figure 7-2. Three Planetary Nebulas
These are three typical planetary nebulas. Most planetary nebulas that we can see in the sky are roughly 20,000 years old. Their diameters can range from a few billion miles to a light year. Intense ultraviolet light from the hot central star can keep the gases glowing for about 50,000 years. (Hale Observatories.)

Figure 7-3. *The Planetary Nebula NGC 7293*
*This beautiful planetary nebula is in the constellation of Aquarius.
Notice the remarkable structure around the inner edge of the
expanding shell of gas. Thousands of small globules have formed
because of instabilities that develop as the shell of gas expands.
These tiny globules shield the outer portions of the nebula from the
central star's ultraviolet light. Hence the globules have a distinct
windblown appearance. (Hale Observatories.)*

About 500 white dwarfs have been discovered around the sky. A good example is the companion star to Sirius, the brightest-appearing star in the sky. Sirius is really a double star and its tiny white dwarf companion can be seen easily in Figure 7-4. The low luminosity and high surface temperature of white dwarfs like Sirius's companion place them in the lower left-hand corner of the Hertz-sprung-Russell diagram, the graveyard. It is where the evolutionary tracks of low-mass stars must end.

An important breakthrough in calculating the evolutionary tracks of dying low-mass stars occurred in 1970. The Polish astrophysicist B. Paczynski succeeded in developing the techniques by which a computer could make the enormous number of computations necessary to follow the rapid changes that occur in a dying star. Our understanding of how red giants become white dwarfs is a direct result of his work.

Three of Paczynski's evolutionary tracks of dying low-mass stars are shown in Figure 7-5. As the outer layers of the star lift away from the burned-out core, the dot representing the star on a H-R diagram moves rapidly across the graph. During this time, the total luminosity remains fairly constant. But as the planetary nebula dissipates, the evolutionary track takes a sharp turn down towards the white dwarf region of the diagram.

Each of Paczynski's evolutionary tracks represents substantial loss of mass. The tracks begin with red giants whose masses are 3 solar masses, 1 ½ solar masses, and $^4/_5$ solar mass, and end with white dwarfs whose masses are 1 $^1/_5$ solar masses, $^4/_5$ solar mass, and $^3/_5$ solar mass respectively. As the table on page 135 indicates, the difference in masses (given in solar masses) of the red giant and white dwarf reveals the mass of the ejected planetary nebula.

Also notice that data are plotted on Figure 7-5. Each dot represents the central star in a planetary nebula whose luminosity and surface temperature have been measured. Each cross represents a white dwarf whose luminosity and surface temperature have been measured. Notice how nicely the data are grouped around Paczynski's evolutionary tracks.

As the years pass by, white dwarfs simply cool off. And as their temperatures drop, their luminosities also decline. On the Hertzsprung-Russell diagram, white dwarfs slowly slip downward toward

Figure 7-4.   A White Dwarf
Sirius, the brightest appearing star in the sky, is actually a double star. Its tiny white dwarf companion is visible in this excellent photograph. White dwarfs are very small —typically the same size as the earth —but they have surface temperatures between 10,000 and 50,000 °K. (Photograph courtesy of R. B. Minton.)

Evolutionary tracks on Figure 8-3	Red giant mass	White dwarf mass	Mass of ejected nebula
Track A	3.0	1.2	1.8
Track B	1.5	0.8	0.7
Track C	0.8	0.6	0.2

the lower-right along *cooling curves.* As shown in Figure 7-6, these cooling curves are almost parallel to the main sequence. In principle, white dwarfs eventually cool off completely. But this cooling process is very slow and it would take trillions of years for a star to approach absolute zero. Since this is much longer than the age of the universe, even the most ancient white dwarfs are still fairly warm.

As a result of the work of people like Hayashi, Iben, and Paczynski, we are now in a position to understand the complete life history of the sun. This history is best displayed on a Hertzsprung-Russell diagram, as shown in Figure 7-7. When the protosun first forms, it is large and cool, and therefore it has a high luminosity and a low temperature just after accreting from the interstellar medium.

In a very short period of time, the sun contracts and its interior becomes hot enough to ignite hydrogen burning. At this stage, the sun arrives at the main sequence and has become a full-fledged star.

After 10 billion years, hydrogen fuel will have been used up at the sun's center. Our star will rapidly become a red giant and will spend one or two billion years in that stage. During this time, the sun's evolutionary track will wander around in the red giant region of the H-R diagram as the sun's temperature and luminosity adjust to the various nuclear processes occurring below its surface.

Finally, near the end of its life, the sun will begin to pulsate. During one of these spasms our doomed star will eject its outer layers. Indeed, the sun could easily shed one-quarter of its mass during these death throes. The result will be a beautiful planetary nebula. During this time, the sun's evolutionary track will race horizontally across the H-R diagram. Then, as the planetary nebula dissipates and the burned-out core contracts, the evolutionary track will

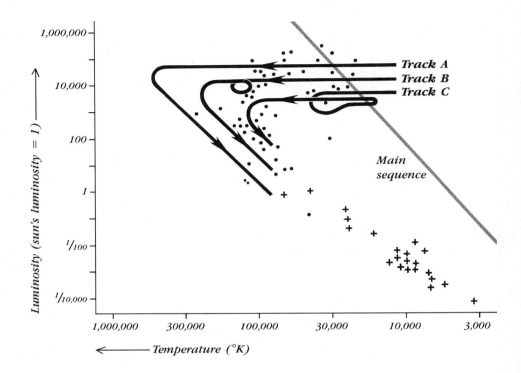

Figure 7-5. Evolution from Red Giants to White Dwarfs
These are the evolutionary tracks of three low-mass red giants as they evolve into white dwarfs. During this metamorphosis, the stars eject large quantities of matter into space (see accompanying table). These changes occur extremely rapidly. Consequently, the dots representing these evolving stars race across the H-R diagram as the stars produce planetary nebulas.

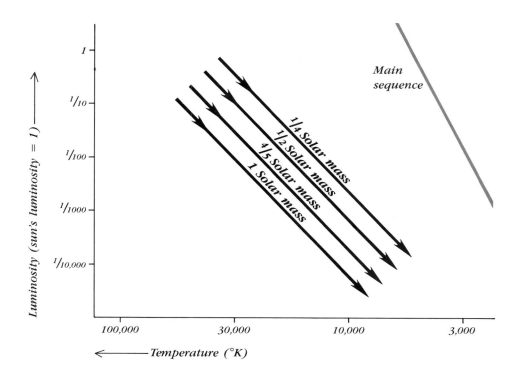

Figure 7-6. Cooling Curves for White Dwarfs
White dwarfs are dead stars. They simply cool off by radiating energy into space. As the surface temperature of a white dwarf declines, its luminosity also decreases. The evolutionary tracks of white dwarfs therefore lead in the direction of decreasing temperature and luminosity, as shown in this graph.

137

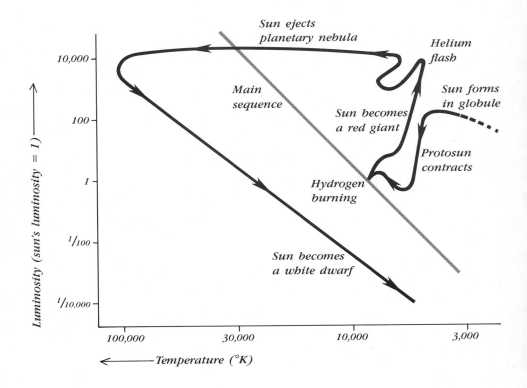

Figure 7-7. *The Sun's Life History*
*The complete evolution of the sun, from birth to death, is shown
on this H-R diagram. The sun first spends 10 billion years as a
main sequence star. About 5 billion years from now, the sun will
become a red giant. After 1 or 2 billion years as a red giant, the
sun will produce a planetary nebula and the burned-out core
will contract to become a white dwarf.*

take a sharp turn downward to the white dwarf region of the H-R diagram. The sun's corpse will be a white dwarf.

Much of our understanding about white dwarfs dates back to the early 1930s and the pioneering work of the Indian (now American) astrophysicist S. Chandrasekhar. Using concepts from quantum mechanics and relativity theory, Chandrasekhar developed the first reliable and accurate theoretical models of white dwarf stars. The surprising result of this work is that there is a maximum mass for white dwarfs. Degenerate electron pressure can only support up to 1.4 solar masses. All white dwarfs must have masses that are smaller than this critical value of 1.4 solar masses. This critical value is called the *Chandrasekhar limit.*

Our dying sun will meet this requirement. In addition, since the vast majority of stars have low masses, most stars must become white dwarfs. Even a star of 2 or 3 solar masses has no difficulty ejecting its outer layers so that the burned-out core contains less than the Chandrasekhar limit.

But what about the most massive stars? Although they are in a distinct minority, there are many stars in the sky that have masses up to 60 or 70 solar masses. What will become of them? Suppose, for example, that a star that has 50 solar masses ejects only 48 solar masses during a supernova. Since the core is more massive than the Chandrasekhar limit, degenerate electron pressure will not be able to hold up the star; it simply cannot become a white dwarf. What happens to the dead stellar core?

Up until the mid-1960s, it was generally supposed that all dying stars eventually became white dwarfs. After all, there seemed to be enough white dwarfs in the sky to account for the corpses of all the stars that should have died by now. So it seemed reasonable to suppose that somehow even the most massive stars would manage to eject enough matter to get below the Chandrasekhar limit. Some astrophysicists have proposed fascinating alternatives to white dwarfs. For example, by the end of the 1930s, J. Robert Oppenheimer and his colleagues had developed some important ideas concerning *neutron stars* and *black holes*. But these proposals were largely ignored as fantasy. After all, aren't there enough white dwarfs around to account for all dead stars? Actually, astronomers simply did not know what else to look for.

139

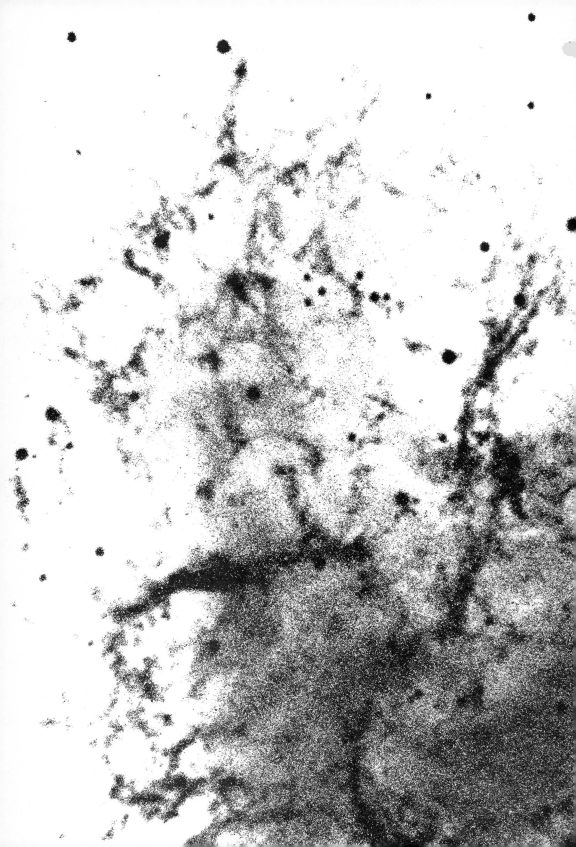

8
Pulsars and Neutron Stars

The prospect of communicating with a superior alien intelligence is one of the most exciting and terrifying possibilities in modern astronomy. Many scientists argue vigorously that advanced intelligent creatures probably inhabit numerous undiscovered planets scattered around our galaxy. The precision and sensitivity of our radio telescopes continue to increase, so it may be just a matter of time before we stumble across an advanced alien race.

It is absurd to suppose that we could imagine what an advanced race of creatures might say to us. Of course, such contemplations are fun; they provide the fantasy fuel for thousands upon thousands of science fiction books. But if the creatures we contact are only slightly ahead of us—say, only two or three million years more advanced than we—then we would face an intellectual gap roughly the same as that that exists between man and the chimpanzee. If we could begin to understand what they are saying, it would be like receiving communications directly from the gods.

Although we could not reasonably expect to guess the contents of a message from advanced alien creatures, we might realistically imagine the overall impact of such a message. After all, there are many clear precedents. Throughout history various people supposedly received suprahuman communications. Moses, Jesus Christ, and Muhammad are excellent examples. The crucial point is this: Many people believed and many today believe that these persons actually received messages from an advanced being. As a result of this widespread belief, the entire course of civilization has been profoundly affected.

Well, we are fresh out of burning bushes. But there are a lot of radio telescopes around. Just as the devoutly religious person does not doubt the authenticity of holy writ, modern scientists could not quarrel with clearly demonstrable data. The reception of a clear and indisputable alien message by an astronomer at a radio telescope is totally analogous to Moses receiving the Ten Commandments from God on Mount Sinai.

Imagine the turmoil and agony that the astronomer would experience! Whom should he tell? How should the announcement be made? Publicly or just to government officials? Perhaps humanity would not be ready to hear the message. Or perhaps the astronomer might be inclined to destroy all records of his observations. Upon

142

Figure 8-1. A Radio Telescope
Radio telescopes are designed to focus, amplify, and record radio waves from space. Using radio telescopes, astronomers have discovered many fascinating objects in the sky, including pulsars and quasars. Someday we might even detect a message from an alien civilization with the use of radio telescopes. (National Radio Astronomy Observatory.)

143

deciphering the message, the astronomer's sanity might be severely affected. The astronomer would clearly face the gravest professional, moral, and personal crisis of his life.

In the fall of 1967, a team of astronomers at the Mullard Radio Astronomy Observatory at the University of Cambridge were forced to confront this crisis head on.

In October of 1967, Jocelyn Bell was making radio observations of the stars with a new array of antennas at Cambridge, England. During one particular observation, she realized that her radio telescope was picking up regular pulses of radio noise. Once every 1.33731109 seconds, a pulse of radio noise was detected. It was soon determined that the timing of the pulses was incredibly regular; only man-made atomic clocks could rival the precision of these pulses. The pulses were therefore assumed to be signals from a secretly launched Soviet satellite.

The idea of a Soviet satellite lasted for only a couple of days. Any satellite, even one on a distant interplanetary voyage, should appear to move with respect to the background stars. But this mysterious pulsating radio source remained fixed among the constellations. The source must be located far beyond the solar system, where no man-made satellites have ever traveled. Since no known natural phenomenon had the precision and regularity of these pulses, it seemed entirely reasonable to consider the possibility that Ms. Bell had discovered a signaling beacon of an advanced alien civilization. In view of the staggering implications of this possibility (as well as the possibility of making complete fools of themselves), the Cambridge radio astronomers held Ms. Bell's discovery in absolute secrecy.

During the next few months, the Cambridge astronomers discussed and debated the issues that must be faced when and if an alien message is finally received. All their deliberations were for naught. By Christmas of 1967, Jocelyn Bell had discovered three more of these pulsating radio sources. Their widely-scattered positions among the stars clearly suggested that Ms. Bell had found a new, extraordinary type of astronomical object, rather than an isolated alien race. With some relief, and perhaps with some disappointment, the "Little Green Men Theory" was abandoned. In its place, astrophysicists began

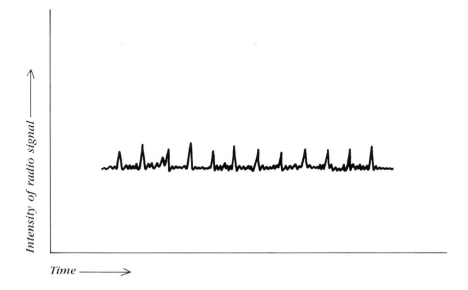

Figure 8-2. A Pulsar Recording

In October of 1967, astronomers at the Mullard Radio Astronomy Observatory detected regular pulses from an object in space. Since that time, nearly 150 pulsars have been found across the sky. Today we realize that a pulsar is a rapidly rotating neutron star that has an intense magnetic field.

145

searching for a natural explanation for these pulsating radio sources, or *pulsars,* as they were soon christened. Today we realize that Jocelyn Bell had actually discovered *neutron stars.*

Think about a massive star that is nearing the end of its life. As we learned in Chapter 6 (see, for example, Figure 6-6), a host of nuclear reactions are occurring in concentric shells inside the star. In the innermost shell, silicon burning at a temperature of 3½ billion degrees Kelvin converts silicon into iron. Gradually, the massive star develops an iron core. The nuclei of iron atoms are the most tightly bound nuclei in nature. They will not burn; they cannot burn. The growth of an iron-rich core at the center of a massive star portends the star's imminent, cataclysmic death.

A star is in serious trouble once it has developed an iron-rich core of roughly 1 solar mass. Since no new thermonuclear reactions are possible, the burned-out core simply contracts and gets hotter and hotter. Finally, the core becomes so dense that the iron nuclei begin breaking up into helium nuclei. In addition, at these incredible densities, the negatively charged electrons get squeezed into the positively charged protons, thereby creating many neutrons. A collection of neutrons occupies a much smaller volume than an equivalent amount of iron nuclei. Consequently, as the electrons and protons combine to form neutrons, the star's core rapidly implodes.

The sudden, catastrophic collapse of a star's core releases an incredible amount of energy. In fact, the amount of energy released during core collapse can be as great as *all* the energy radiated by the star over its entire preceding life! As the energy from core collapse surges outward, the star is completely ripped apart. In one of nature's most violent cataclysms, the star's luminosity suddenly increases a billion-fold. For a few days, this single star can be as bright as the entire galaxy in which it resides. The star has become a *supernova.*

Occasionally astronomers are lucky enough to see a supernova in the process of exploding. However, these supernovas are usually located in distant galaxies and therefore can only be viewed through powerful telescopes. Nearby supernovas that can be seen with the naked eye are much more rare. During the past 1,000 years, only four nearby supernovas have been seen. They occurred in 1006 (in Lupus), in 1054 (in Taurus), in 1572 (in Cassiopeia), and in 1604

Figure 8-3. *A Distant Supernova*
During 1959, a supernova exploded in this galaxy. These are views of the galaxy before (top) and after (bottom) the explosion. At maximum brilliancy, a supernova can be as bright as an entire galaxy. (Lick Observatory Photograph.)

147

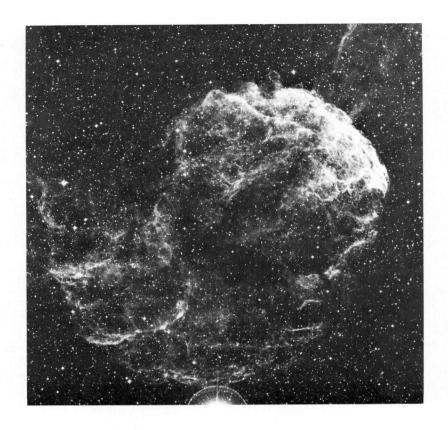

Figure 8-4. A Supernova Remnant
*This beautiful filamentary nebula, called IC 443, is a supernova
remnant. Tens of thousands of years ago, an ancient massive star
ended its life by becoming a supernova. The spectacular detonation
blasted enormous quantities of stellar material out into space.
(Hale Observatories.)*

(in Serpens). These supernovas were so bright that they could be seen during broad daylight. It has been more than three centuries since the last nearby supernova occurred that could be seen with the naked eye and many astronomers feel that we are overdue for a whopper.

Although it is unusual to see a supernova in the process of exploding, astronomers often find evidence of ancient supernovas scattered around the sky. These objects, called *supernova remnants,* are the outer portions of the star that were violently ejected during the supernova outburst. Excellent examples of supernova remnants are shown in Figures 6-7 and 8-4. Some of the most beautiful and delicate nebulas in the sky are supernova remnants.

As we have discussed, the onset of a supernova explosion is triggered by the collapse of a dying star's burned-out core. Explosive thermonuclear processes are ignited above the collapsing core and the star is torn apart in a spectacular cosmic detonation. A dying massive star can eject most of its matter during a supernova explosion. In doing so, the star enriches interstellar space with numerous heavy elements that had been manufactured inside the star's thermonuclear furnaces. But what happens to the star's burned-out, neutron-rich core?

During core-collapse, protons and electrons are crushed together to form neutrons. Finally, when 1 to 2 solar masses of this neutron-rich material are squeezed into a sphere roughly 20 miles in diameter, the neutrons become degenerate. Recall that degenerate electron pressure can stop the gravitational contraction of dead low-mass stars. White dwarfs are dead low-mass stars that are supported by degenerate electron pressure. In exactly the same sense, *degenerate neutron pressure* can stop the collapse of a star's burned-out core and support a slightly more massive dead star. This kind of a stellar corpse is called a *neutron star.*

In a white dwarf, up to 1.4 solar masses of burned-out stellar matter have been squeezed into a sphere roughly 10,000 miles in diameter. Inside the star, the density is typically 1,000 tons per cubic inch. At this density, the electrons simply cannot be squeezed any closer; they have become degenerate. Degenerate electron pressure supports the star.

In a neutron star, up to 3 solar masses of burned-out stellar matter have been squeezed into a sphere roughly 20 miles in diameter. Inside the star, electrons and protons have combined to form neutrons and the density is typically 10 billion tons per cubic inch. At this enormous density, the neutrons simply cannot be squeezed any closer; they have become degenerate. Degenerate neutron pressure now supports the star.

The neutron was first discovered in laboratory experiments in 1932 by the British nuclear physicist James Chadwick. A year later, in 1933, Caltech astronomers Fritz Zwicky and Walter Baade proposed the possible existence of neutron stars. For the next 35 years, most astronomers either ignored this prophetic proposal completely or regarded it as fantasy and nonsense. Then in 1968 an object was discovered in the sky that could only be interpreted as a neutron star.

"On the day of Ch'uh-Ch'iu in the fifth month of the first year of the Chih-Ho Period, a 'guest star' appeared at the southeast of Thien-Kuan, measuring several inches. After more than a year, it faded away." With these words, a Chinese historian recorded the appearance of a supernova on July 4, 1054, in the constellation of Taurus, the bull. If we turn our telescopes toward the location given in this ancient Chinese record, we see the famous Crab Nebula. As shown in Figure 8-5, the Crab Nebula is a beautiful supernova remnant. In the fall of 1968, astronomers at the National Radio Astronomy Observatory in West Virginia discovered a pulsar directly at the center of the Crab Nebula.

The discovery of the Crab Pulsar (also called NP 0532) came at a very critical time. Jocelyn Bell and her colleagues had recently announced their discovery of the first four pulsars, and astronomers around the world were proposing numerous (sometimes absurd) ideas to account for the regular pulses of radio noise. The discovery of the Crab Pulsar promptly put most of these ideas in the garbage can.

The Crab Pulsar has two obvious, important characteristics. First of all, it is located in a supernova remnant. This proves that pulsars are left over from the deaths of stars. Secondly, the Crab Pulsar is the fastest known pulsar. It pulses once every 0.03309 seconds, nearly 30 times a second. It was immediately clear to everyone that white dwarfs *cannot* pulse, ring, vibrate, or rotate 30 times per

Figure 8-5. The Crab Nebula (also called M1 or NGC 1952)
*This remarkable nebula is a supernova remnant. The supernova
explosion was observed by Chinese astronomers 900 years ago. The
Crab Nebula is 6,500 light years away and it measures about 8
light years in diameter. A pulsar (called NP 0532) is at the center of
this nebula. (Lick Observatory Photograph.)*

151

second. Therefore, the burned-out stellar object left over at the center of the Crab Nebula *cannot* be a white dwarf.

Up until 1968, white dwarfs had reigned supreme as the only generally accepted, recognized stellar corpse. But with the discovery of the Crab Pulsar, astronomers realized that a second kind of object could be created during the deaths of stars. All of the fantastical ideas about neutron stars were promptly revived.

In thinking about neutron stars, astronomers realized that these stellar corpses have three characteristics that directly tie them to pulsars. First of all, as we have seen, neutron stars are small. The diameter of a neutron star is typically 20 miles. But in addition, the neutron star must be rotating rapidly. A basic law of physics, called the conservation of angular momentum, dictates that if any slowly rotating object contracts, its rate of rotation must gain speed. This is why an ice skater doing a pirouette gains speed as she pulls in her arms. For exactly the same reason, as the slowly rotating core of a dying star collapses, its rate of rotation speeds up. The resulting neutron star must be rotating rapidly—roughly once a second, or even faster.

Finally, neutron stars are expected to have intense magnetic fields. In the case of an ordinary star like the sun, the intensity of the magnetic field is fairly weak. After all, the magnetic field is spread over millions upon millions of square miles around the star's surface. But if a star collapses to a diameter of 20 miles, the magnetic field becomes very compressed because it is then confined to a surface area of only a few dozen square miles. The result is a billion-fold increase in the strength of the magnetic field.

As first noted by Franco Pacini and Thomas Gold at Cornell, neutron stars should be small and rapidly rotating and should have intense magnetic fields. It is furthermore reasonable to expect that the magnetic field of a neutron star is inclined at some angle to its axis of rotation. As in the case of the earth or the sun, there is no reason why the magnetic axis and rotation axis have to be parallel. The oblique orientation of the magnetic field of a neutron star to its axis of rotation is shown schematically in Figure 8-6. Notice that the vertical orientation of the magnetic field at the north and south magnetic poles provides corridors along which charged particles can

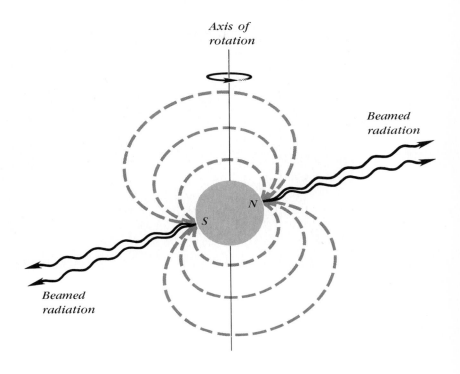

Figure 8-6. The Model of A Pulsar

Pulsars are rapidly rotating neutron stars that have intense magnetic fields. Particles and radiation stream outward along narrow beams from the north and south magnetic poles. As the neutron star rotates, the beams sweep across the sky like beams from a lighthouse beacon. A person who stands in the direction of the beam can see periodic pulses of radiation.

153

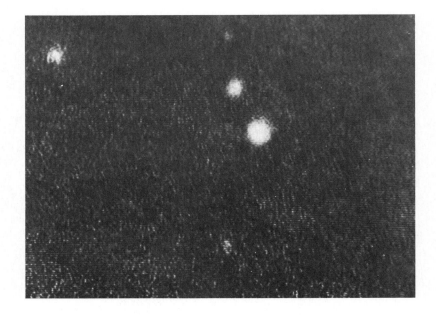

Figure 8-7.   A Visible Pulsar
In 1969, astronomers succeeded in detecting visible flashes from the pulsar at the center of the Crab Nebula. The star turns on and

stream outward away from the neutron star. This continuous flow of particles along these corridors produces two pencil thin beams of coherent radiation. As the neutron star rotates, the beams of radiation sweep across the sky. If someone happens to be looking in the appropriate direction, he will see a flash of light or pulse of radiation each time the beam sweeps by.

This is how pulsars work: in principle, they are the same as an old-fashioned lighthouse. Every time we detect a pulse of radio noise, we know that a beam of radiation from a rotating neutron star has swept across our line of sight.

154

off 30 times each second. The "on" and "off" views are shown in these two photographs. (Lick Observatory Photograph.)

Upon hearing about this "oblique-rotator" model of pulsars, a team of astronomers at the Steward Observatory in Arizona attempted to detect visible flashes from pulsars. After all, why shouldn't visible pulses be seen along with the radio pulses? Using electronic equipment (because the human eye cannot see 30 flashes per second), the astronomers examined the stars at the center of the Crab Nebula. In January 1969, they were delighted to find that one particular star at the center of the nebula is actually blinking on and off 30 times each second. Both "on" and "off" views of the Crab Pulsar are shown in Figure 8-7.

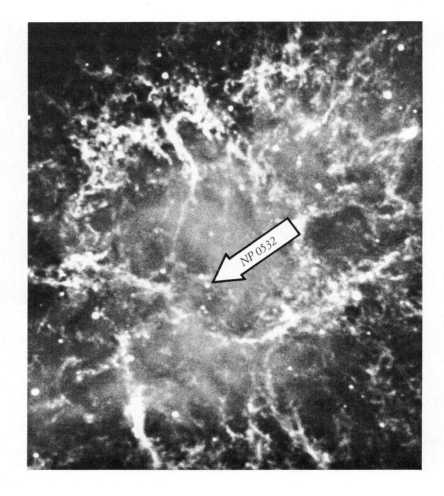

Figure 8-8. A Neutron Star
The discovery of a visible pulsar at the center of the Crab Nebula permitted the identification of a neutron star. This tiny, rapidly rotating neutron star is the powerful source of particles and radiation that have kept the nebula glowing for nine centuries. (Lick Observatory Photograph.)

In spite of many bold efforts, astronomers have succeeded in detecting visible flashes from only one other pulsar. All other pulsars, of which nearly 150 are known, have been observed only at radio wavelengths.

The discovery of pulsars has had a profound effect on the course of astronomy. For decades, neutron stars were dismissed as science fiction and ridiculous fantasy. But now, as shown in Figure 8-8, we actually see them in the sky. As astronomers began examining the properties of neutron stars, they started to wonder whether the most bizarre object ever predicted by science — the black hole — might also someday be found in space.

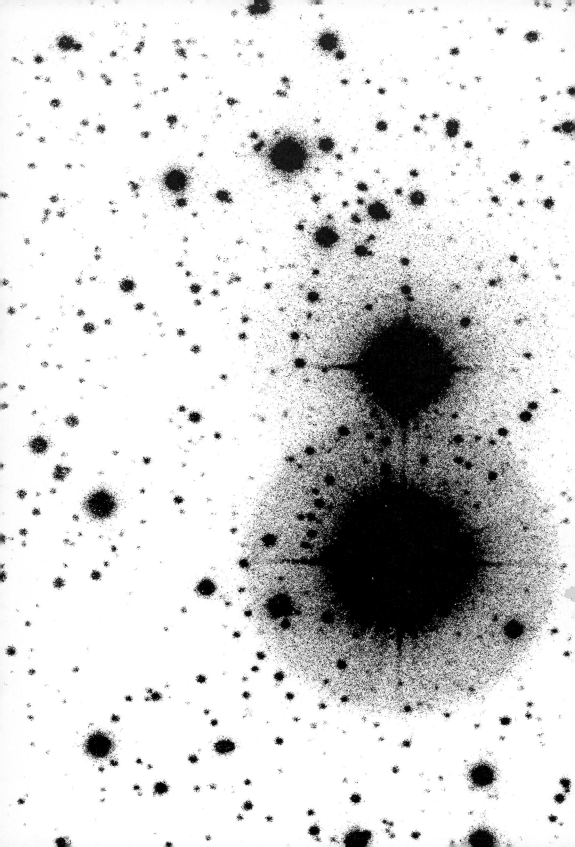

9
Black Holes in Space-Time

The universe is home to an incredible array of objects and phenomena. There are electrons and galaxies, people and planets, watermelons and asteroids. Yet, in spite of this diversity and complexity, physicists find only *four* forces in nature. All of the physical properties of everything around us, all of the physical phenomena and interactions we observe in the universe, are a direct result of these four forces.

At the smallest scale in nature, there are strong and weak nuclear forces. Strong nuclear force holds the nuclei of atoms together. It is the most powerful force in nature, and yet it operates only over very tiny distances. Strong nuclear force is the "nuclear glue" that binds the protons and neutrons together in the nuclei of atoms.

Weak nuclear force is also a short-range force. It comes into play in nuclear interactions involving radioactive decay. But due to their short-range nature, both strong and weak nuclear forces are important only over subatomic distances. Although these forces dictate the nuclear structure and properties of matter, they have no influence over large distances, and consequently do not play any direct role in the universe at the astronomical scale.

The electric interaction between charged particles and the magnetic interaction between magnets or current-carrying wires are really different aspects of the *same* force. This is the third force in nature: the electromagnetic force. Its strength lies between those of strong and weak nuclear forces. The strong nuclear force is typically 100 times more powerful than the electromagnetic force. And the electromagnetic force is roughly a million times more powerful than the weak nuclear force.

Electromagnetic force dominates the world of atoms and molecules. Negatively charged electrons stay in orbit about positively charged nuclei because of the electromagnetic force they exert on each other. This same force allows atoms to combine and form molecules. As a result, millions of molecules can be formed from only 92 naturally occurring chemical elements. In addition to giving an incredible diversity to the universe, some of these molecules have the extraordinary ability to manufacture exact replicas of themselves. This phenomenon is called life.

Unlike the strong and weak nuclear forces, electromagnetic force *is* long range. But once again, we would not expect this force to play an influential role at the astronomical scale. The reason is that for every north magnetic pole there is a south magnetic pole. Similarly, the vast majority of physical situations are electrically neutral: for every positive electric charge there is a negative electric charge. Consequently, over large distances and dimensions, the effects cancel out to zero.

The fourth force is the most familiar in nature — gravity. We encounter it at every moment of our daily lives. Yet gravity, that universal attraction of matter for matter, is by far the weakest force in nature. To understand how very weak gravity is, imagine two electrons floating near each other in outer space. These two electrons exert an electromagnetic force on each other that tries to push them apart (the adage "opposites attract and likes repel" may have come from early experiments with electricity and magnetism). But they also contain matter, and all matter has gravity. Thus, the electrons exert a gravitational force on each other that tries to pull them together. The electromagnetic force is a million trillion trillion trillion times stronger than the gravitational force between the two electrons.

Gravity is a long-range force just like electromagnetic force. But unlike electricity and magnetism, there is no "negative gravity" or repulsive gravity. All gravity is attractive; if you pile more and more matter into a particular volume of space, the intensity of gravity around that region always increases. So, although it is genuinely the weakest force in nature, gravity can have a strong effect in regions where large quantities of matter are piled up — such as in stars or galaxies.

For these reasons, gravity is the only important force at the astronomical scale. It keeps the moon in orbit about the earth. It holds the solar system together. It dominates the interactions between stars and between galaxies. It even dictates the entire past and future of the universe as a whole.

Everything we know about gravity dates back to the brilliant, pioneering work of Sir Isaac Newton in the 1660s. Newton's goal was to understand the motions of the planets around the sun. In examining the orbits of the planets, we see fundamental physical laws revealed in their purest and simplest forms, unhampered by friction or air resistance encountered in laboratory experiments. Newton's precise formulation of the force of gravity, and the concomitant birth of classical physics, were destined to come from the heavens.

Newton's *Universal Law of Gravitation* has proved eminently successful as a description of nature's weakest force. Using Newton's law, astronomers of the eighteenth and nineteenth centuries found that they could calculate the orbits of planets and satellites, of comets and asteroids, with incredible precision. Every time (with the lone exception of one very small detail involving the orbit of Mercury, the innermost planet), Newton's law of gravity emerged triumphant.

Shortly after the turn of the twentieth century, a young German physicist with a deep personal conviction reexamined Newtonian gravitation and did not like what he saw. His name was Albert Einstein, and he felt that the laws of physics should not depend on the peculiarities of particular observers. Einstein felt that if we are truly discovering fundamental laws of nature, the way in which we write down these laws should be totally independent of our particular location or state of motion in the universe. This was not true of Newton's work, so Einstein set about the business of developing a new description of gravity. The final result was the *General Theory of Relativity.*

The General Theory of Relativity is a theory of gravity. It tells us how gravity works. But in sharp contrast to Newton's theory, general relativity never treats gravity as a "force." Instead, gravity is described entirely in terms of the curvature of space and time. The stronger the gravitational field, the greater the curvature of spacetime.

To help themselves to visualize the warping effect of gravity, physicists sometimes draw diagrams such as Figure 9-1. This illustration, called an embedding diagram, shows the geometry of space around a massive object like the sun or a star. Notice that space is flat far from the source of gravity. But progressively closer to it are regions of higher and higher curvature.

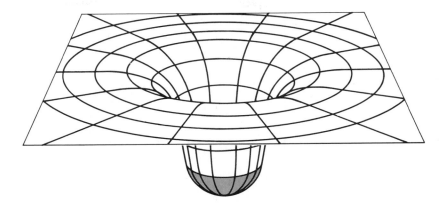

Figure 9-1.   The Gravitational Curvature of Space
According to general relativity, gravity curves space-time. This drawing, called an embedding diagram, shows how space is curved around a massive object such as the sun or a star. The shaded region indicates the location of the star. The greatest curvature (and thus the most intense part of the star's gravitational field) is found immediately above the star's surface. Far from the star, where gravity is very weak, space-time is almost perfectly flat.

Figure 9-2. The Orbit of Mercury
The innermost planet in the solar system exhibits a small but unusual motion that cannot be calculated or explained from classical Newtonian physics. As Mercury tries to go around the sun in an elliptical orbit, the orbit itself rotates. Consequently, the planet traces a rosette figure (highly exaggerated in this drawing). Einstein's general theory of relativity completely accounts for this anomalous motion.

To complement this geometrical approach to the subject of gravitation, Einstein argued that objects moving through curved space-time simply travel along the shortest possible paths. Far from any sources of gravitation, where space-time is flat, objects move along perfectly straight lines. But near a source of gravity, where space-time is curved, objects move along curved orbits. These orbits, the curved space-time counterparts of straight lines, are called *geodesics*.

After formulating his general theory of relativity, Einstein naturally wanted to see whether it predicted the right answers in various situations. The most obvious test was the solar system and he calculated the orbits of the planets around the sun. Except for the planet Mercury, Einstein's relativistic approach gave exactly the same answers as Newton's classical approach. But recall that Newtonian gravitation had failed to predict Mercury's orbit with complete accuracy. The innermost planet in the solar system exhibits a very tiny motion, called precession, that cannot be explained or calculated from Newton's law of gravity. In December of 1915, Einstein completed his calculations of Mercury's orbit. He found that general relativity gave *exactly* the right answer, including a complete accounting for the anomalous precession.

In addition to giving a complete explanation of the orbit of Mercury, Einstein also predicted that light rays should be deflected by gravity. Quite simply, when a light ray travels through a gravitational field, it is moving through curved space-time. Consequently, as shown in Figure 9-3, the light ray must travel along an arc rather than a perfectly straight line. This is a purely relativistic effect; no one had ever considered the possibility that gravity could deflect light rays.

In calculating the gravitational deflection of light, Einstein realized that this phenomenon could only be noticed near the sun. The sun is the most massive object in the solar system. It possesses the strongest gravitational field and therefore is surrounded by the greatest curvature of space-time in the solar system. And yet, by relativistic standards, the sun's gravity is quite weak. Only for a light ray grazing the sun's surface would the deflection be large enough to detect.

165

A total solar eclipse occurred on May 29, 1919. During the precious few minutes when the moon blocked out the blinding disc of the sun, astronomers photographed the stars around the sun's edge. Careful measurement revealed that the stellar images had been deflected slightly from their usual positions just as Einstein had predicted.

The precession of Mercury's orbit and the gravitational deflection of light were great triumphs for Einstein's general theory of relativity. For over two centuries, Newton's law of gravity had endured as the cornerstone in the foundations of classical physics. And then suddenly, in 1916, there was a radically new theory of gravity that worked even better.

This triumph of relativistic gravitation over classical gravitation caused a great deal of excitement that turned out to be short-lived. It was soon realized that, except for tiny details of Mercury's orbit or light rays grazing the sun's surface, Einstein's theory gave precisely the same answer as Newton's theory around the solar system. This is because space-time is nearly flat in our solar system and around our galaxy. In the nearly flat space-time of a weak gravitational field, Einstein's theory and Newton's theory are equivalent. No one could realistically imagine a place in the universe where the warping of space-time is severe, where the intensity of gravity is so inconceivably enormous that relativistic effects dominate everything. So why bother with the complicated concepts and difficult mathematics of general relativity? Interest in general relativity rapidly waned.

In 1939, J. Robert Oppenheimer and his colleagues published two papers. In the first, Oppenheimer and Volkoff showed that neutron stars should exist. In the second, Oppenheimer and Snyder argued that massive dying stars should become black holes. This second paper was a bold and prophetic attempt to show that there might be places in the universe dominated by the effects of highly warped space-time. For three decades, conservative astronomers dismissed these pioneering works as nothing more than an amusing trip into fantasyland.

The discovery of pulsars in the late 1960s dramatically changed this situation. It was soon realized that neutron stars are the

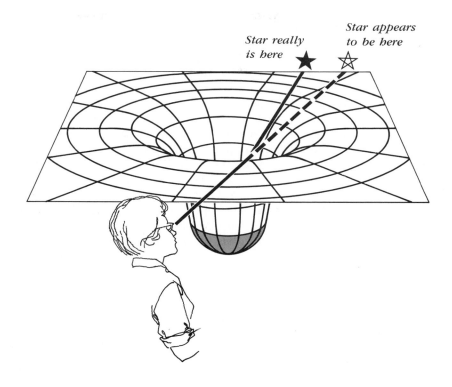

Figure 9-3. The Gravitational Deflection of Light
*Light rays moving through curved space-time must be deflected
from their usual straight paths. Einstein therefore predicted that
stars seen near the sun at the time of a total solar eclipse will be
shifted slightly from their usual positions. This deflection was first
observed during an eclipse in 1919. It proved that Einstein had a
better and more accurate description of gravity than Newton.*

logical explanation for pulsars. If one of these fantasyland objects can actually exist in nature, why not the other? The 1960s therefore saw a sudden resurgence of interest in general relativity as astrophysicists around the world began doing theoretical calculations about black holes.

A black hole is one of the three types of stellar corpses. But unlike white dwarfs or neutron stars, the black hole represents the complete triumph of gravity over matter. Recall that in white dwarfs, the weight of all the burned-out stellar matter is supported by degenerate electron pressure. Similarly, neutron stars are supported by degenerate neutron pressure. By 1970, calculations on the structure of neutron stars showed that degenerate neutron pressure cannot hold up more than 3 solar masses of burned-out stellar matter. A dead star of more than 3 solar masses is doomed to ultimate gravitational implosion. As trillions upon trillions of tons of burned-out matter press inward from all sides, the star simply gets smaller and smaller. As the massive dead star shrinks in size, the intensity of gravity above the star gets stronger and stronger. In terms of general relativity, this means that the curvature of space-time around the star becomes greater and greater. Finally, when enough matter has been squeezed into a small enough volume, space and time fold over themselves and the star disappears from the universe! What is left is called a black hole.

To appreciate some of the properties of a black hole, imagine a massive dying star. Suppose that after a supernova explosion, the remaining dead star contains more than 3 solar masses so that it could not become a white dwarf or neutron star. Before the onset of gravitational collapse, while the star's matter is still spread out over a fairly large volume, the intensity of gravity around the star is still fairly weak. In terms of general relativity this means that space-time around the star is only mildly curved. As shown in Figure 9-4 (View A), light rays leaving any point on the star will travel along nearly straight lines.

But as the gravitational collapse proceeds, the star's matter is squeezed into a smaller and smaller volume. Consequently, the intensity of gravity above the dead star becomes stronger and therefore the curvature of space-time becomes more pronounced. As shown in Figure 9-4 (Views B and C), this causes light rays to be bent more and more from their usual straight-line paths. Finally, gravity becomes so strong that *all* light rays are bent back down towards the star, as shown schematically in Figure 9-4 (View D). At this stage in the collapse, we say that the star has fallen inside its *event horizon*. Since light cannot get out, nothing else can get out. The star has literally vanished from the universe.

Once a dead massive star has collapsed inside its event horizon, the star permanently disappears from the universe. By way of an example, the event horizon of a black hole of 10 solar masses is 60 kilometers (or 37 miles) in diameter. Thus, as soon as the star collapses to a size less than 60 kilometers in diameter, a black hole is formed. But even inside the event horizon there is still nothing to hold up all the dead stellar matter. The star continues to get smaller and smaller until it is totally crushed out of existence at a single point called the *singularity*.

As shown in Figure 9-5, the structure of a black hole consists of an event horizon surrounding a singularity. At the event horizon, gravity is so strong that nothing — not even light — can escape from the hole. Inside the event horizon, the strength of gravity and the curvature of space-time continue to increase right up to the singularity at the center of the hole. Indeed, at the singularity there is infinite curvature of space-time.

It is important to realize that a black hole is *empty*. There is nothing there. All the matter of the massive dead star has been crushed out of existence at the singularity. All that remains is a highly warped region of space and time. Quite literally, the black hole represents the bizarre triumph of gravity — the weakest force in nature — over everything, including the fabric of space and time.

As you may well imagine, finding black holes around our galaxy poses some difficult problems for the astronomer. Since black holes do not emit light, it is impossible to see them. The only hope we have of finding a black hole comes from the possibility of observing the influence of the hole's enormous gravitational field on something that we can see. For example, we might stand a chance of

169

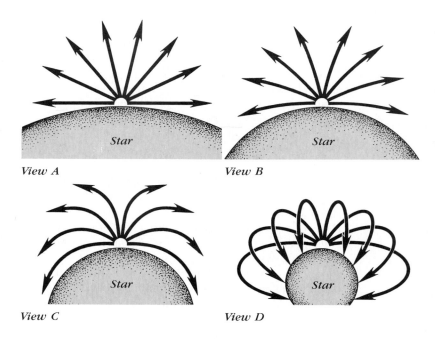

Figure 9-4. Light Rays from A Collapsing Star

As a massive dead star collapses, the intensity of gravity above the star gets stronger and stronger. This means that, as the collapse proceeds, light rays leaving the star's surface are bent through increasingly larger angles. View A shows the star just before the onset of collapse; space-time around the star is only slightly curved and light rays leave the star along nearly straight lines. View B and View C show the star during collapse; as the curvature of space-time around the star increases, light rays become increasingly deflected from their usual straight paths. This deflection eventually becomes so severe that, as shown in View D, all light rays are bent back down to the collapsing star's surface. At this stage the star has fallen inside its own event horizon and a black hole has formed.

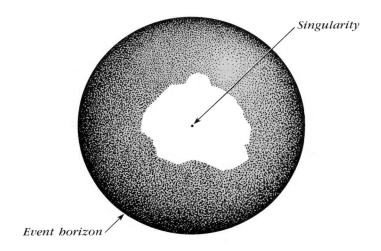

Figure 9-5. The Structure of A Black Hole

A black hole consists of an event horizon surrounding a singularity. At the event horizon, gravity is so strong that nothing —not even light —can escape from the hole. Inside the event horizon, the intensity of gravity and curvature of space-time continue to increase right up to the singularity. At the singularity, there is infinite curvature of space-time.

seeing matter or gas falling into a black hole. Just before the material takes the final plunge down the drain in space-time, radiation might be emitted that announces the existence of the hole. This would amount to an "indirect" discovery since the black hole itself is never observed. The astronomer must therefore take great care in interpreting his observations to prove that all other explanations of the data are ruled out. Only then could the astronomer announce the discovery of a black hole with confidence.

On December 12, 1970, the United States launched a small satellite from a platform in the ocean off the coast of Kenya. In recognition of the hospitality of the Kenyan people, the satellite was christened *Uhuru,* the Swahili word meaning "Freedom." The launch date was the seventh anniversary of Kenyan independence.

Uhuru (formerly called Explorer 42) was the first in a series of small astronomical satellites. It was designed to detect X rays from astronomical sources. Prior to the launch of Uhuru, astronomers had discovered a few X-ray sources in the sky during brief rocket flights. Unfortunately, these small rockets spent only a few tantalizing moments above the obscuring effects of the atmosphere before plunging back down to earth. But with the launch of Uhuru, astronomers could make continuous X-ray observations of the entire sky without interruption.

Uhuru, shown in Figure 9-6, consists of two banks of X-ray telescopes back to back. As the satellite slowly rotates, the X-ray telescopes scan the sky. If an object that emits X rays passes across the field of view, X-ray detectors are activated and signals are transmitted down to earth. From knowing the orientation of the X-ray telescopes at the time of the signals, the astronomer can deduce the location of the source in the sky. After almost three years of flawless operation, 161 X-ray sources had been observed and listed in the *Third Uhuru Catalogue.* Uhuru ushered in the age of X-ray astronomy. For the first time, we could really see what the X-ray sky looks like.

Early in 1971, attention was focused on one particular X-ray object called Cygnus X-1 (also called 3U 1956+35 in the *Third Uhuru Catalogue*). This strong source of X rays had been discovered

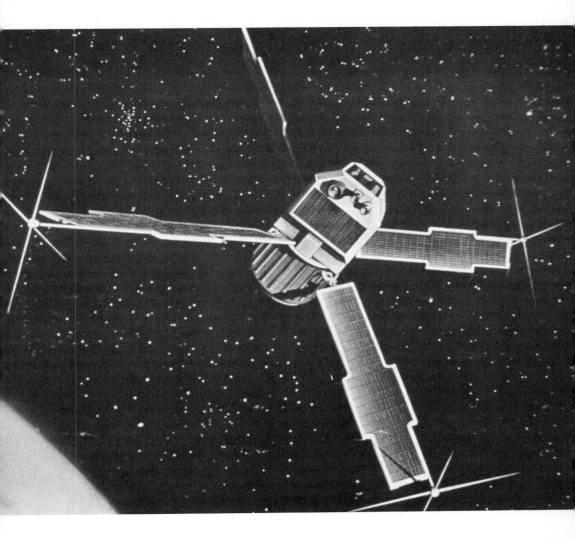

Figure 9-6. Uhuru

This small satellite contains two banks of X-ray detectors back to back. As the satellite slowly rotates, the detectors scan the sky for X-ray sources in space. This satellite gave astronomers their first complete look at the sky in X rays. (NASA.)

173

in 1965 during one of those brief rocket flights. As its name suggests, it was the first X-ray source detected in the constellation of Cygnus, the swan. Observations from Uhuru during March and April of 1971 revealed that X-ray emissions from Cygnus X-1 were changing dramatically. During these changes, a radio source appeared in the sky at the location of Cygnus X-1. Using the giant 140-foot radio telescope at the National Radio Astronomy Observatory, astronomers were able to pin point the source. The radio source (and, presumably, also the X-ray source) coincided with a star called HDE 226868. This star (the 226,868th in the extension to the famous Henry Draper catalogue) is indicated in Figure 9-7.

HDE 226868 is a rather ordinary, hot, massive, bright, bluish star. Such stars do not emit any appreciable amounts of X rays or radio waves. So the visible star itself could not be Cygnus X-1. But Cygnus X-1 clearly had to be nearby.

As you may recall from Chapter 3, roughly half of the stars you see in the sky are actually double stars—two stars in orbit about each other. During the spring and summer of 1972, observations of HDE 226868 revealed that it was in a binary star system. The orbital period is 5.6 days and the companion star is invisible. The logical conclusion seemed to be that the invisible companion is, in fact, Cygnus X-1.

Everything we know about Cygnus X-1 can be fully explained and understood from the viewpoint that Cygnus X-1 is a black hole orbiting the visible star HDE 226868. The giant, hot, visible star is constantly emitting matter in the form of a stellar wind. As this matter encounters the gravitational field of the black hole, the matter is captured into orbit about the hole. As shown in Figure 9-8, the captured gases form a huge disc around the hole, like a giant version of the rings around Saturn. It is believed that the disc is roughly 3½ million kilometers (or about 2 million miles) in diameter.

We can understand the emission of X rays that we observe as Cygnus X-1 by examining how the gases in the disc actually orbit the black hole. Just as Mercury goes about the sun faster than Pluto, the inner portions of the disc revolve about the hole more rapidly than the outer portions. As the slowly moving outer portions rub against

Figure 9-7.   HDE 226868
Team work by radio astronomers and X-ray astronomers in 1971 gave a very accurate determination of the position of Cygnus X-1: it has the same location as the star HDE 226868 (indicated by the arrow). Observations in 1972 showed that HDE 226868 is one member of a double-star system. The invisible companion is Cygnus X-1. (Courtesy of J. Kristian, Hale Observatories.)

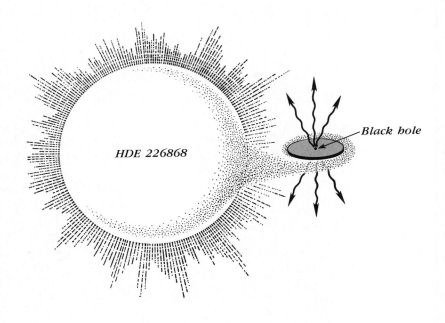

HDE 226868

Black hole

Figure 9-8. The Cygnus X-1 Black Hole
The stellar wind from HDE 226868 pours matter onto a huge disc about its black hole companion. The infalling gases are heated to enormous temperatures as they spiral towards the black hole. At the inner edge of the disc, just above the black hole, the gases are so hot that they emit vast quantities of X rays.

the rapidly rotating inner portions, friction develops. This causes the gases to heat up and to spiral inward toward the hole. As the gases spiral closer and closer to the hole, the temperature gets higher and higher. Finally, only a few hundred kilometers above the hole, the temperature has climbed to several million degrees. For example, at a distance of 100 kilometers (60 miles) from the hole, the temperature has risen to an incredible 10 million degrees Kelvin. At these tremendous temperatures, the disc becomes transparent and radiation can pour out. Wien's law tells us that anything at several million degrees emits enormous quantities of X rays. Indeed, the innermost portions of this disc emit 10,000 times more energy in X rays than the sun emits in ordinary visible light.

This interpretation of Cygnus X-1 was greatly strengthened in the mid-1970s as astronomers detected the motions of HDE 226868 about its unseen companion. Recall from Chapter 3 that by observing stars in a binary, astronomers can (ideally) determine the masses of the stars. In the case of the system containing Cygnus X-1 and HDE 226868, it is now clear that Cygnus X-1 contains at least 8 solar masses. This is far too massive to be a white dwarf or a neutron star. The only remaining alternative is a black hole.

Epilogue

With the discovery of a black hole, one of the most bizarre and fantastical predictions of modern science has been verified. Indeed, there are incredible places in the universe where gravity overwhelms all other forces in nature and where space and time fold over themselves around a collapsing, massive star. Since the discovery of a black hole, concepts and topics formerly relegated to science fiction stories are now commonly discussed at scientific meetings.

It may turn out that black holes play a more important role in the universe than anyone had ever realized. This seems especially true with regard to quasars and exploding galaxies. The more astronomers learn about galaxies and quasars, the more they are inclined to turn to black holes to explain the enormous energies and violent processes we observe in the cosmos. Under appropriate circumstances, the large amount of energy tied up in the gravitational field of a black hole can be used to produce exceptionally high luminosities or to power the vigorous ejection of large quantities of gas and plasma. Indeed, all quasars and active galaxies may contain supermassive black holes. Even more incredibly, numerous very tiny black holes could have been created during the violent and chaotic birth of the universe. These tiny black holes may have acted as the seeds around which galaxies first condensed billions upon billions of years ago.

Of all the forces in nature, gravity dominates the behavior of the universe on the astronomical scale. And the best theory of gravity we have, the general theory of relativity, clearly predicts the existence of black holes. Yet, in an unexpected turn of fate, it was re-

179

cently discovered that black holes do not obey the laws of physics — the self-same laws that allowed black holes to exist in the first place!

The whole purpose of physics is to understand *what* happens *where* and *when*. The "where" and "when" means that all the laws of physics are formulated against a well-defined background of space and time. But at the singularity at the center of a black hole, there is infinite curvature of space-time. Therefore, at the singularity, space and time are so completely twisted and entangled that they are not separated, well-defined entities. Since space and time cease to exist at the singularity, the laws of physics — so firmly rooted in space and time — need not be obeyed. The singularity can do anything it wants without any regard to the laws of nature!

This remarkable theoretical discovery by Stephen W. Hawking, called the Randomicity Principle, is of no consequence as long as the vicious singularity is perpetually shielded from the outside universe by an event horizon. As long as an event horizon separates the singularity from the outside universe, any chaotic events at the singularity cannot get out of the black hole to influence the rest of the universe. But if tiny black holes were created during the "Big Bang," then the situation would be very different. Chaotic events at the singularity can through quantum mechanics tunnel out of these tiny black holes to the outside universe. Indeed, in the mid-1970s, Stephen Hawking proved that, at the quantum level, black holes are indistinguishable from white holes that randomly spew matter and energy into the universe. If these tiny black holes really exist, then science is severely limited in what it can determine about the universe. As scientists try to understand the game, the cards and the players come and go and change in a totally random, chaotic fashion.

For Further Reading

For a survey of modern astronomy, the reader might wish to consult an elementary college text. Six excellent college texts are listed below.

Exploration of the Universe (Third Edition), George O. Abell (Holt, Rinehart and Winston, 1975).

Exploring the Cosmos (Second Edition), Louis Berman and J. C. Evans (Little, Brown & Co., 1977).

Astronomy: The Structure of the Universe, William J. Kaufmann (Macmillan; 1977).

The Universe Unfolding, Ivan R. King (W. H. Freeman & Co., 1976).

Contemporary Astronomy, Jay M. Pasachoff (W. B. Sanders, 1977).

Astronomy: The Evolving Universe, Michael Zeilik (Harper & Row, 1976).

To pursue selected topics in stellar astronomy and stellar evolution, the reader might want to consult the following books.

Atoms, Stars and Nebulae, L. H. Aller (Harvard University Press, 1971).

The Milky Way (Fourth Edition), B. J. Bok and P. F. Bok (Harvard University Press, 1974).

New Frontiers in Astronomy, ed: O. Gingerich (W. H. Freeman & Co., 1975).

The Cosmic Frontiers of General Relativity, W. J. Kaufmann (Little, Brown & Co., 1977).

During recent years, a number of exceptional articles written for the layperson have appeared in popular journals. Some of the best are listed below, and they are grouped according to subject.

Stellar Birth and Young Stars

"The Birth of Stars," Bart J. Bok, *Scientific American,* vol. 227, no. 2, pp. 49-61 (August 1972).

"Bok Globules," Robert L. Dickman, *Scientific American*, vol. 236, no. 6, pp. 66-81 (June 1977).

"The Youngest Stars," George H. Herbig, *Scientific American*, vol. 217, no. 2, pp. 30-36 (August 1967).

"Interstellar Clouds and Molecular Hydrogen," Michael Jura, *American Scientist*, vol. 65, no. 4, pp. 446-454 (July-August 1977).

The Sun

"The Case of the Missing Sunspots," John A. Eddy, *Scientific American*, vol. 236, no. 5, pp. 80-92 (May 1977).

"Waves in the Solar Wind," J. T. Gosling and A. J. Hundhausen, *Scientific American*, vol. 236, no. 3, pp. 36-43 (March 1977).

"The Sun," E. N. Parker, *Scientific American*, vol. 233, no. 3, pp. 42-50 (September 1975).

"The Solar Corona," Jay M. Pasachoff, *Scientific American*, vol. 229, no. 4, pp. 68-79 (October 1973).

Mature and Old Stars

"Globular-Cluster Stars," Icko Iben, Jr., *Scientific American*, vol. 223, no. 1, pp. 26-39 (July 1970).

"Pulsating Stars," John R. Percy, *Scientific American*, vol. 232, no. 6, pp. 66-75 (June 1975).

"Evolution of Red Giant Stars," Allen V. Sweigart, *Physics Today*, vol. 29, no. 1, pp. 25-32 (January 1976).

"Cosmology: Man's Place in the Universe," Virginia Trimble, *American Scientist*, vol. 65, no. 1, pp. 76-86 (January-February 1977).

Supernovas

"X Rays from Supernova Remnants," Philip A. Charles and J. Leonard Culhane, *Scientific American*, vol. 233, no. 6, pp. 38-46 (December 1975).

"Supernova Remnants," Paul Gorenstein and Wallace Tucker, *Scientific American*, vol. 225, no. 1, pp. 74-83 (July 1971).

"The Gum Nebula," Stephen P. Maran, *Scientific American*, vol. 225, no. 6, pp. 21-29 (December 1971).

"Historical Supernovas," F. Richard Stephenson and David H. Clark, *Scientific American*, vol. 234, no. 6, pp. 100-107 (June 1976).

Stellar Corpses

"X-Ray Stars in Globular Clusters," George W. Clark, *Scientific American*, vol. 237, no. 4, pp. 42-55 (October 1977).

"X-Ray Emitting Double Stars," Herbert Gursky and Edward P. J. van den Heuvel, *Scientific American*, vol. 232, no. 3, pp. 24-35 (March 1975).

"The Nature of Pulsars," Jeremiah P. Ostriker, *Scientific American*, vol. 224, no. 1, pp. 48-60 (January 1971).

"Black Holes," Roger Penrose, *Scientific American*, vol. 226, no. 5, pp. 38-46 (May 1972).

"The Search for Black Holes," Kip S. Thorne, *Scientific American*, vol. 231, no. 6, pp. 32-43, (December 1974).

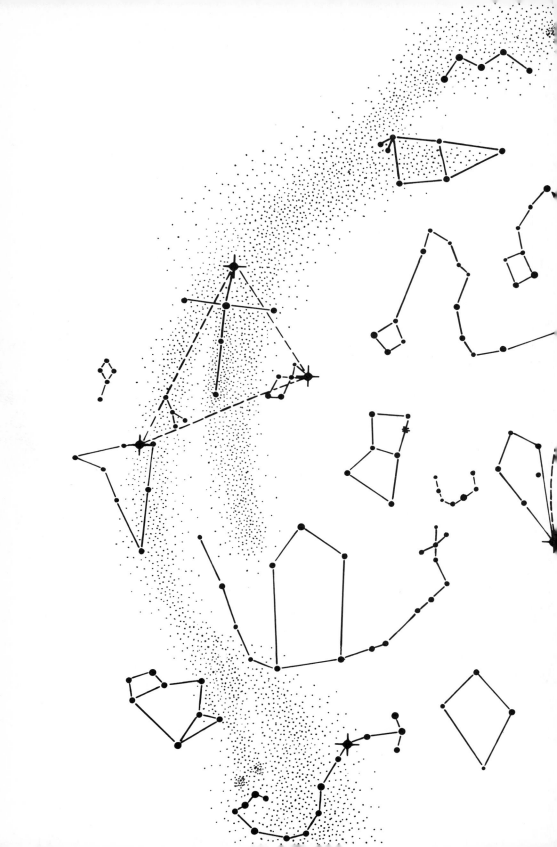

Monthly
Star Charts

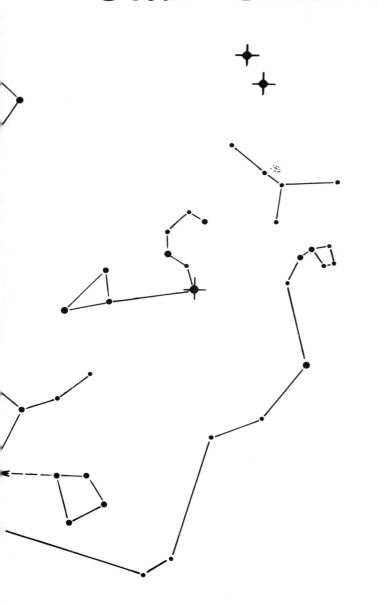

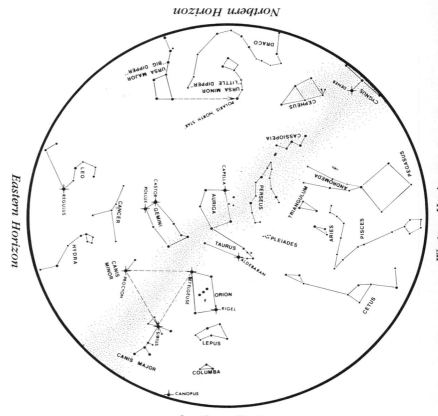

The night sky in January

Chart time *(local standard)*
10 pm, first of the month
9 pm, middle of the month
8 pm, last of the month

To use *Hold chart vertically and turn it so
the direction you are facing shows at the bottom.*

From Griffith Observer magazine, Griffith Observatory

186

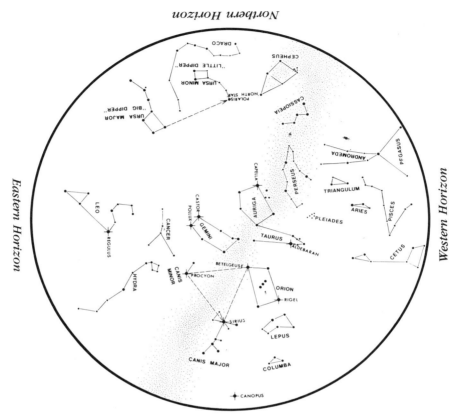

The night sky in February

Chart time (*local standard*)
10 pm, first of the month
9 pm, middle of the month
8 pm, last of the month

To use   *Hold chart vertically and turn it so
the direction you are facing shows at the bottom.*

From Griffith Observer magazine, Griffith Observatory

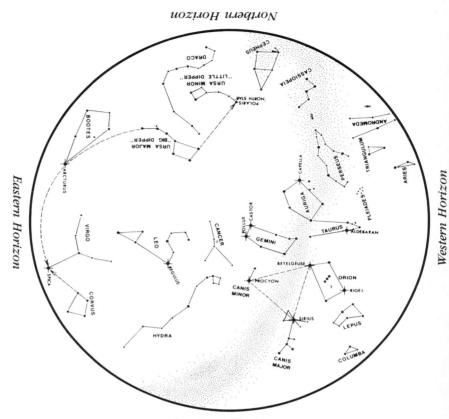

Northern Horizon

Eastern Horizon

Western Horizon

Southern Horizon

The night sky in March

Chart time *(local standard)*
10 pm, first of the month
9 pm, middle of the month
8 pm, last of the month

To use *Hold chart vertically and turn it so*
the direction you are facing shows at the bottom.

From Griffith Observer magazine, Griffith

188

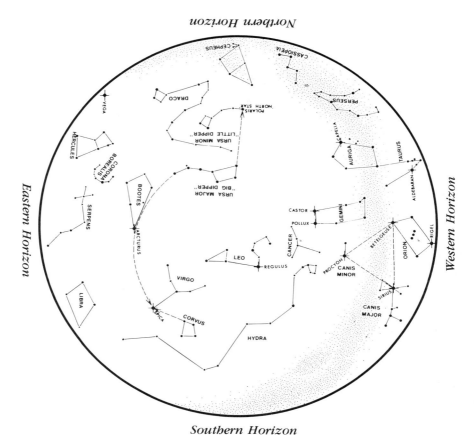

Southern Horizon

The night sky in April

Chart time (*local standard*)
10 pm, first of the month
9 pm, middle of the month
8 pm, last of the month

To use Hold chart vertically and turn it so
the direction you are facing shows at the bottom.

From Griffith Observer magazine, Griffith Observatory

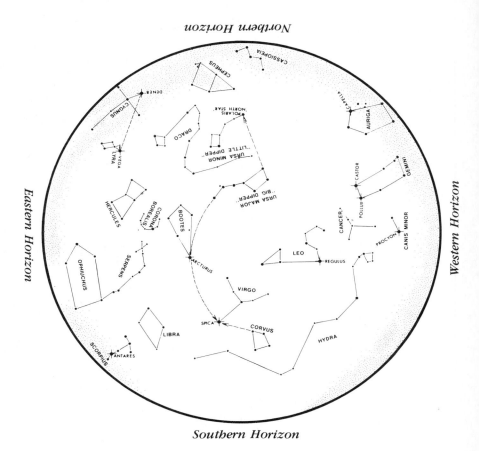

Southern Horizon

The night sky in May

Chart time *(local standard)*
10 pm, first of the month
9 pm, middle of the month
8 pm, last of the month

To use *Hold chart vertically and turn it so*
the direction you are facing shows at the bottom.

From Griffith Observer magazine, Griffith Observatory

190

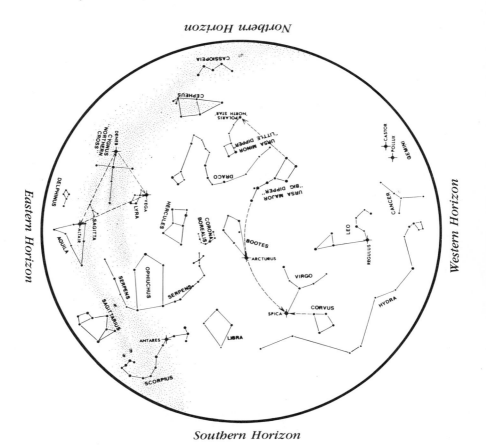

The night sky in June

Chart time *(local standard)*
10 pm, first of the month
9 pm, middle of the month
8 pm, last of the month

To use *Hold chart vertically and turn it so the direction you are facing shows at the bottom.*

From Griffith Observer magazine, Griffith Observatory

191

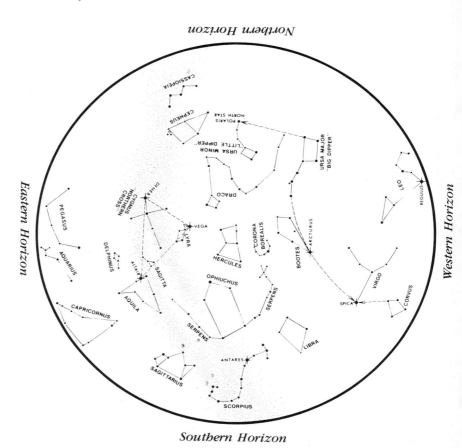

Southern Horizon

The night sky in July

Chart time *(local standard)*
10 pm, first of the month
9 pm, middle of the month
8 pm, last of the month

To use *Hold chart vertically and turn it so
the direction you are facing shows at the bottom.*

From Griffith Observer magazine, Griffith Observatory

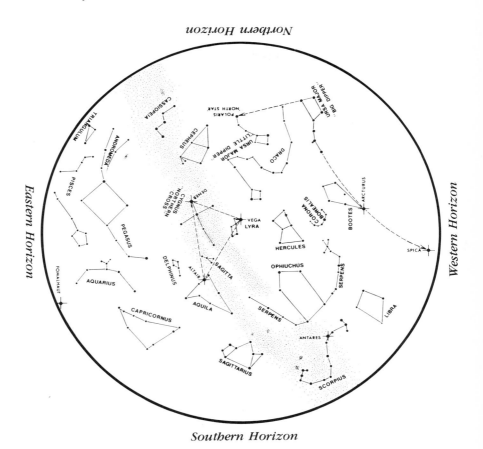

The night sky in August

Chart time (local standard)
10 pm, first of the month
 9 pm, middle of the month
 8 pm, last of the month

To use   Hold chart vertically and turn it so
the direction you are facing shows at the bottom.

From Griffith Observer magazine, Griffith Observatory

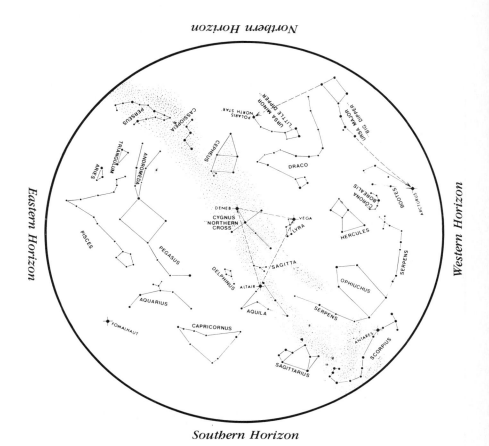

The night sky in September

Chart time *(local standard)*
10 pm, first of the month
9 pm, middle of the month
8 pm, last of the month

To use *Hold chart vertically and turn it so
the direction you are facing shows at the bottom.*

From Griffith Observer magazine, Griffith Observatory

194

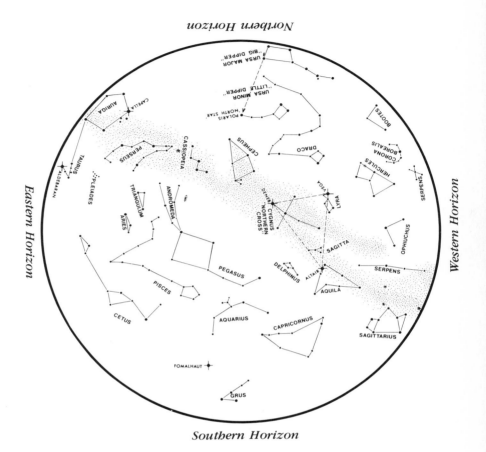

The night sky in October

Chart time *(local standard)*
10 pm, first of the month
9 pm, middle of the month
8 pm, last of the month

To use *Hold chart vertically and turn it so
the direction you are facing shows at the bottom.*

From Griffith Observer magazine, Griffith Observatory

195

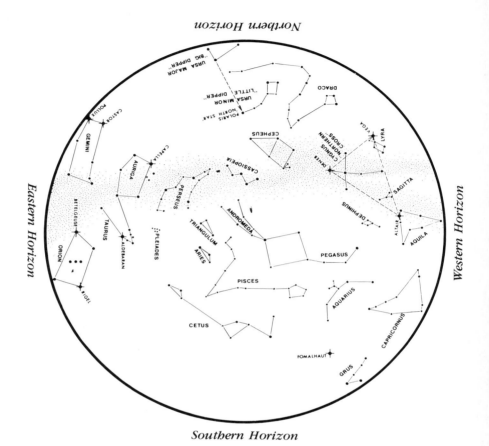

Southern Horizon

The night sky in November

Chart time *(local standard)*
10 pm, first of the month
9 pm, middle of the month
8 pm, last of the month

To use *Hold chart vertically and turn it so
the direction you are facing shows at the bottom.*

From Griffith Observer magazine, Griffith Observatory

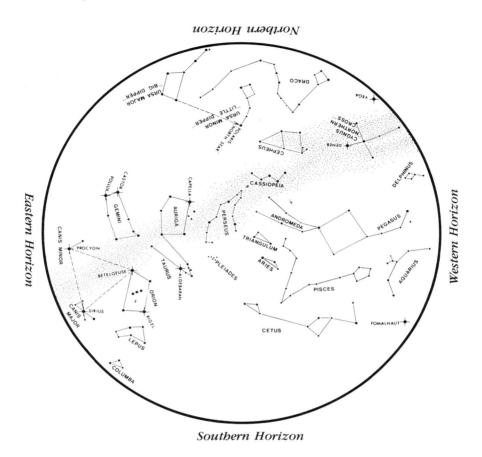

The night sky in December

Chart time *(local standard)*
10 pm, first of the month
9 pm, middle of the month
8 pm, last of the month

To use *Hold chart vertically and turn it so
the direction you are facing shows at the bottom.*

From Griffith Observer magazine, Griffith Observatory

Index